Power PRACTICE™

MW01706655

Fractions and Decimals

3–4

Written by
Vicky Shiotsu

Editor: Carla Hamaguchi
Illustrator: Jenny Campbell
Production: Carrie Rickmond
Cover Designer: Barbara Peterson
Art Director: Moonhee Pak
Project Manager: Collene Dobelmann
Project Director: Betsy Morris

Table of Contents

Introduction

Fractions and Decimals 3–4 contains ready-to-use activity pages to provide students with skill practice. The fun activities can be used to supplement and enhance what you are already teaching in your classroom. Give an activity page to students as independent class work, or send the pages home as homework to reinforce skills taught in class. An answer key is included at the end of the book.

This book provides activities that will directly assist students in practicing basic skills and concepts. The structure of the book enhances students' learning and enables them to meet new challenges with confidence. The book is divided into two sections: the first section features fractions while the second presents decimals. The pages in the book introduce skills in an orderly progression to ensure students' success. Many of the pages also provide problem-solving activities that allow students to apply what they learned and to practice critical thinking skills. In addition, two review pages at the end of each section let you evaluate students' learning.

Students will receive reinforcement in the following skills:

- comparing fractions
- ordering fractions
- adding and subtracting fractions
- changing improper fractions to mixed numbers
- adding mixed numbers
- matching fractions and decimals
- comparing decimals
- ordering decimals
- rounding decimals
- adding and subtracting decimals

Use *Fractions and Decimals Grades 3–4* to reinforce or extend concepts and skills. "Recharge" skill review with the ready-to-go activities in this book, and give students the power to succeed!

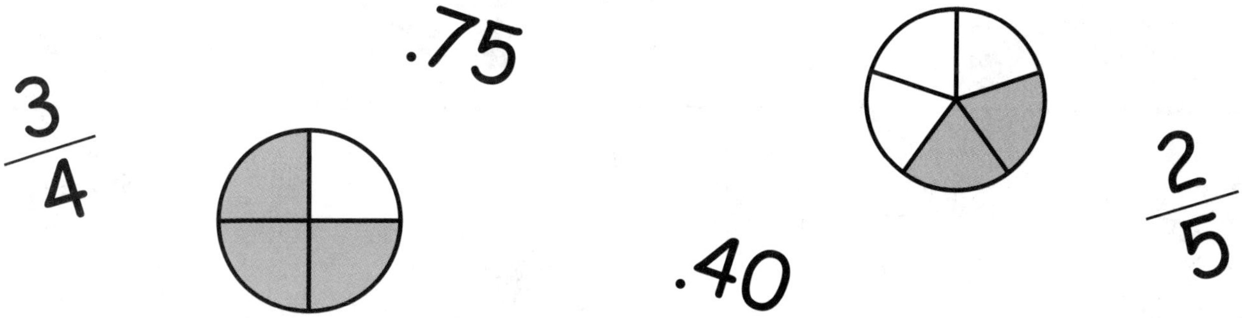

Name ______________________________ Date ______________

Identifying Parts of a Whole

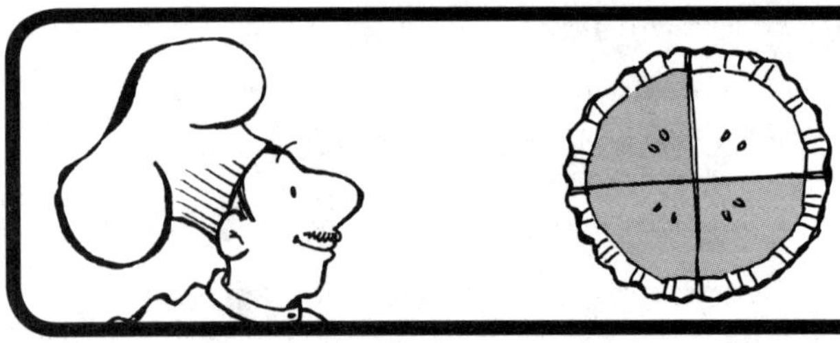

3 ← The **numerator** tells how many parts you are comparing to the whole.

4 ← The **denominator** tells the total number of equal parts of a whole.

For each picture, write a fraction that tells what part is shaded and what part is not shaded.

A

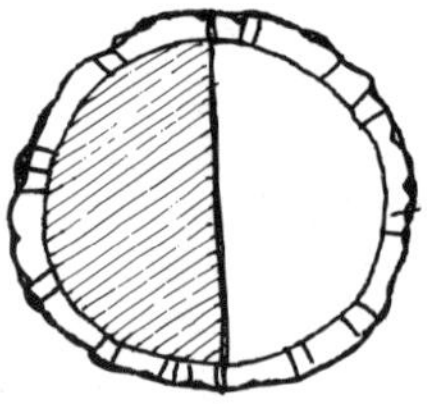

Shaded | Not Shaded

B

Shaded | Not Shaded

C

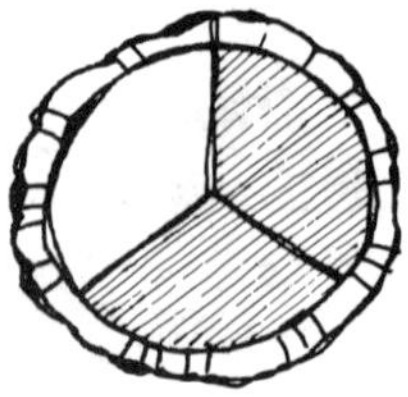

Shaded | Not Shaded

D

Shaded | Not Shaded

E

Shaded | Not Shaded

F

Shaded | Not Shaded

G

Shaded | Not Shaded

H

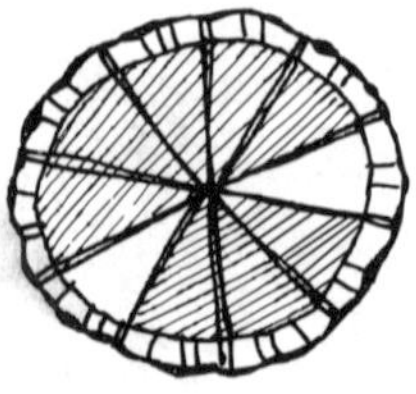

Shaded | Not Shaded

Name ______________________________ Date ______________

Identifying Parts of a Group

$\frac{2}{5}$

The **numerator** tells what part you are comparing to a whole group.

The **denominator** tells the total number of parts in a whole group.

Write a fraction that tells what part is shaded.

A

B

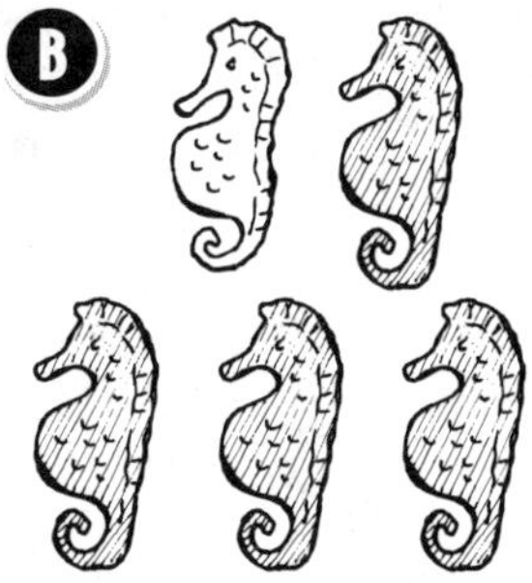

C

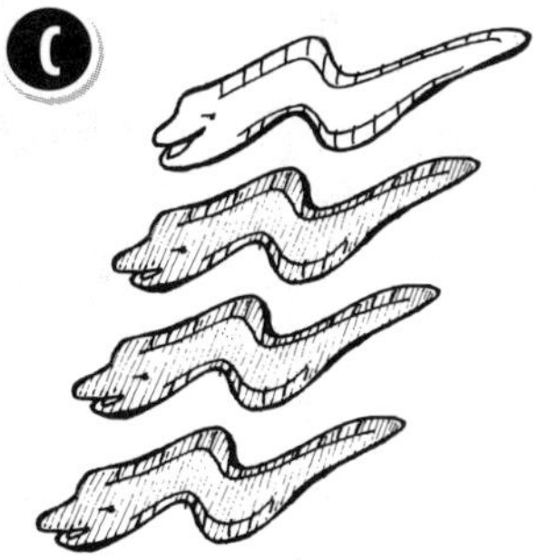

D

E

F

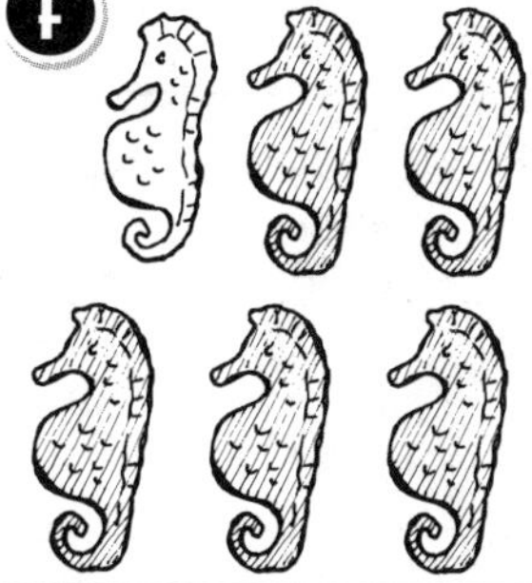

G

H

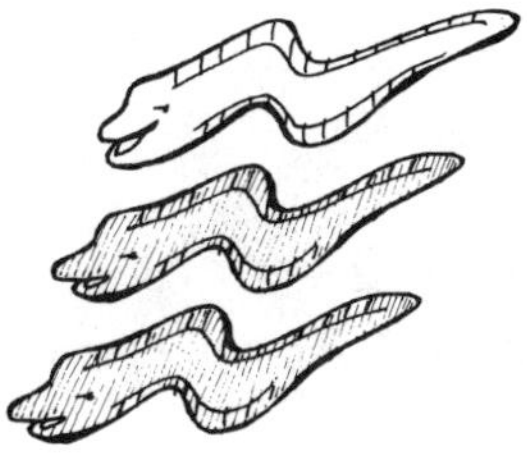

I

J Draw 5 fish. Color ⅕ of the fish red.
Color ⅗ of them yellow.

What fraction of the fish are red or yellow? ____________

What fraction of the fish are not colored? ____________

Name ______________________________ Date ______________

Comparing Fractions—Like Denominators

Color the bars to make them match the fractions in each pair. Circle the greater fraction.

A $\frac{2}{3}$

$\frac{1}{3}$

B $\frac{3}{4}$

$\frac{2}{4}$

C $\frac{2}{5}$

$\frac{4}{5}$

D $\frac{5}{8}$

$\frac{7}{8}$

Compare the fractions. Write > or < in the circles.

E $\frac{4}{6}$ ◯ $\frac{5}{6}$ $\frac{3}{5}$ ◯ $\frac{1}{5}$ $\frac{6}{8}$ ◯ $\frac{1}{8}$ $\frac{2}{9}$ ◯ $\frac{3}{9}$

F $\frac{5}{7}$ ◯ $\frac{6}{7}$ $\frac{2}{10}$ ◯ $\frac{9}{10}$ $\frac{4}{12}$ ◯ $\frac{7}{12}$ $\frac{4}{8}$ ◯ $\frac{3}{8}$

G Jim and Kate each had a strip of paper 24 inches long.
Jim used ⅜ of his strip. Kate used ⅝ of her strip. Who used more paper? ________

H Brett and Kylie each had a piece of ribbon 18 inches long.
Brett used 2⁄6 of his ribbon for an art project.
Kylie used 3⁄6 of her ribbon. Who had more ribbon left? ______________

I If two fractions have the same denominator, how can you tell which one is greater?______________________________

Name ______________________________ Date ______________

Comparing Fractions—Like Numerators

When comparing fractions that have the same numerator, look at the denominators. The fraction with the smaller denominator is larger.

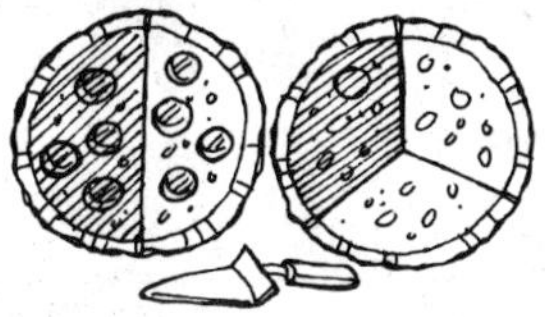

$\frac{1}{2} > \frac{1}{3}$

Compare the fractions. Write > or < in the circles.

A	$\frac{1}{5}$ ○ $\frac{1}{2}$	$\frac{2}{7}$ ○ $\frac{2}{6}$	$\frac{3}{4}$ ○ $\frac{3}{5}$	$\frac{7}{8}$ ○ $\frac{7}{10}$
B	$\frac{2}{3}$ ○ $\frac{2}{6}$	$\frac{5}{8}$ ○ $\frac{5}{10}$	$\frac{4}{9}$ ○ $\frac{4}{5}$	$\frac{9}{12}$ ○ $\frac{9}{10}$
C	$\frac{7}{12}$ ○ $\frac{7}{11}$	$\frac{6}{9}$ ○ $\frac{6}{8}$	$\frac{2}{4}$ ○ $\frac{2}{6}$	$\frac{8}{10}$ ○ $\frac{8}{12}$

D A pepperoni pizza and a mushroom pizza were the same size.
Kris ate $\frac{2}{10}$ of the pepperoni pizza.
Lee ate $\frac{2}{8}$ of the mushroom pizza.

Who ate more pizza? ____________________

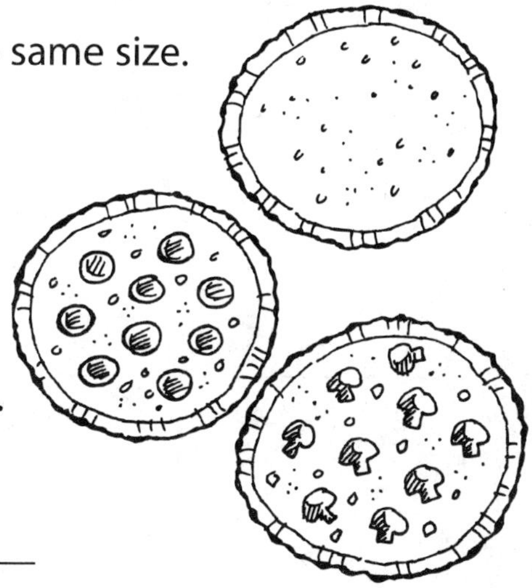

E Evan and Lori were making pizzas.
Evan sprinkled $\frac{2}{3}$ cup of grated cheese on his pizza.
Lori sprinkled $\frac{2}{4}$ cup of cheese on her pizza.

Who used more cheese? ____________________

F Mr. Burton's class did a survey to see which pizza topping was the most popular.
One fourth of the students chose cheese.
One half of the students chose pepperoni.
One sixth of the students chose ham.

Which topping was the most popular? ____________________

Of the three toppings, which one was the least popular? ____________________

Name ______________________________ Date ______________

Ordering Fractions

Write the fractions in order from the least to the greatest.

A $\frac{4}{5}$ $\frac{1}{5}$ $\frac{3}{5}$ $\frac{2}{5}$

____

B $\frac{3}{6}$ $\frac{5}{6}$ $\frac{4}{6}$ $\frac{2}{6}$

____

C $\frac{7}{10}$ $\frac{2}{10}$ $\frac{5}{10}$ $\frac{9}{10}$

____

D $\frac{6}{8}$ $\frac{3}{8}$ $\frac{1}{8}$ $\frac{5}{8}$

____

E $\frac{1}{9}$ $\frac{1}{6}$ $\frac{1}{5}$ $\frac{1}{8}$

____

F $\frac{2}{7}$ $\frac{2}{4}$ $\frac{2}{10}$ $\frac{2}{3}$

____

G $\frac{5}{12}$ $\frac{5}{8}$ $\frac{5}{7}$ $\frac{5}{11}$

____

H $\frac{3}{4}$ $\frac{3}{12}$ $\frac{3}{10}$ $\frac{3}{8}$

____

I $\frac{4}{5}$ $\frac{4}{8}$ $\frac{4}{6}$ $\frac{4}{9}$

____

J $\frac{6}{8}$ $\frac{6}{10}$ $\frac{6}{12}$ $\frac{6}{7}$

____

Name ______________________________ Date ______________

Equivalent Fractions

Fractions that describe the same amount are called **equivalent fractions**.

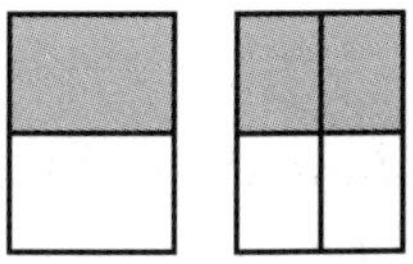

$\frac{1}{2}$ $\frac{2}{4}$

Use the pictures below to help you write equivalent fractions.

A

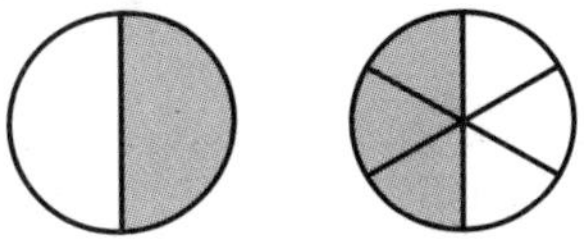

$\frac{1}{2} = \frac{\_\_}{6}$

B

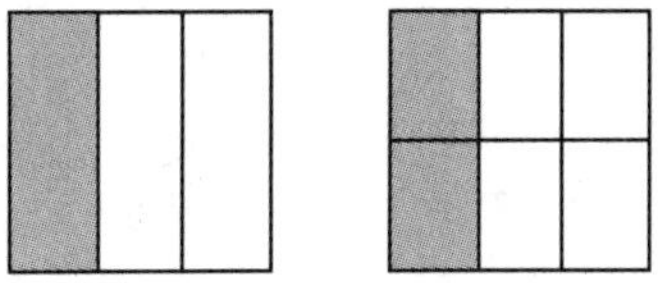

$\frac{1}{3} = \frac{\_\_}{6}$

C

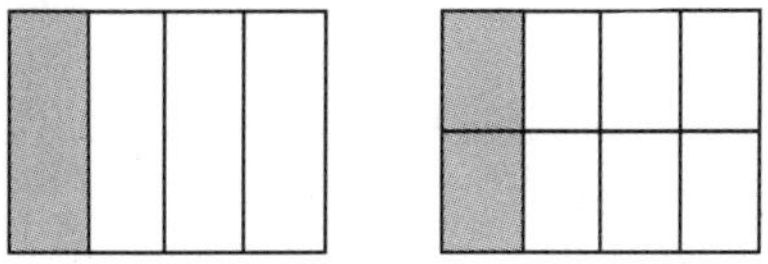

$\frac{1}{4} = \frac{\_\_}{8}$

D

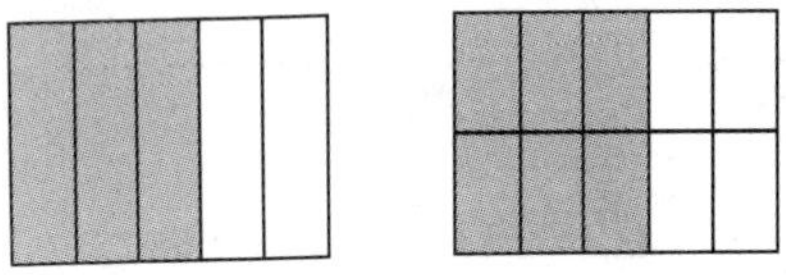

$\frac{3}{5} = \frac{\_\_}{10}$

E

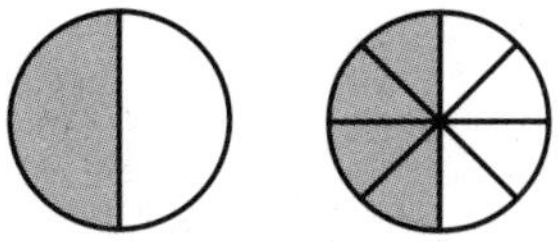

$\frac{1}{2} = \frac{\_\_}{8}$

F

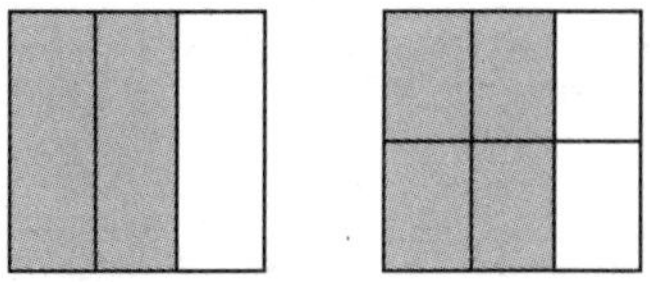

$\frac{2}{3} = \frac{\_\_}{6}$

Use the fraction bars to help you write equivalent fractions.

G $\frac{2}{4} = \frac{\_\_}{8}$ $\frac{3}{4} = \frac{\_\_}{8}$

H $\frac{1}{3} = \frac{\_\_}{9}$ $\frac{2}{3} = \frac{\_\_}{9}$

I $\frac{2}{2} = \frac{\_\_}{4}$ $\frac{3}{3} = \frac{\_\_}{6}$

<table>
<tr><td colspan="8">1 whole</td></tr>
<tr><td colspan="4">½</td><td colspan="4">½</td></tr>
<tr><td colspan="2">¼</td><td colspan="2">¼</td><td colspan="2">¼</td><td colspan="2">¼</td></tr>
<tr><td>⅛</td><td>⅛</td><td>⅛</td><td>⅛</td><td>⅛</td><td>⅛</td><td>⅛</td><td>⅛</td></tr>
</table>

<table>
<tr><td colspan="18">1 whole</td></tr>
<tr><td colspan="6">⅓</td><td colspan="6">⅓</td><td colspan="6">⅓</td></tr>
<tr><td colspan="3">⅙</td><td colspan="3">⅙</td><td colspan="3">⅙</td><td colspan="3">⅙</td><td colspan="3">⅙</td><td colspan="3">⅙</td></tr>
<tr><td colspan="2">⅑</td><td colspan="2">⅑</td><td colspan="2">⅑</td><td colspan="2">⅑</td><td colspan="2">⅑</td><td colspan="2">⅑</td><td colspan="2">⅑</td><td colspan="2">⅑</td><td colspan="2">⅑</td></tr>
</table>

Name ______________________________ Date ______________

Finding Equivalent Fractions

$$\frac{\mathbf{1} \times 2}{\mathbf{2} \times 2} = \frac{\mathbf{2}}{\mathbf{4}} \qquad \frac{\mathbf{3} \div 3}{\mathbf{6} \div 3} = \frac{\mathbf{1}}{\mathbf{2}}$$

Multiply to find equivalent fractions.

A $\frac{1 \times 2}{3 \times 2} = \frac{\_\_}{6}$ $\frac{1 \times 4}{2 \times 4} = \frac{\_\_}{8}$ $\frac{1 \times 3}{4 \times 3} = \frac{\_\_}{12}$

Divide to find equivalent fractions.

B $\frac{4 \div 2}{10 \div 2} = \frac{\_\_}{5}$ $\frac{4 \div 4}{8 \div 4} = \frac{\_\_}{2}$ $\frac{3 \div 3}{9 \div 3} = \frac{\_\_}{3}$

Write equivalent fractions.

C $\frac{1}{2} = \frac{\_\_}{10}$ $\frac{1}{2} = \frac{\_\_}{12}$ $\frac{1}{3} = \frac{\_\_}{12}$

D $\frac{6}{8} = \frac{\_\_}{4}$ $\frac{3}{12} = \frac{\_\_}{4}$ $\frac{2}{8} = \frac{\_\_}{4}$

E $\frac{9}{12} = \frac{\_\_}{4}$ $\frac{2}{3} = \frac{\_\_}{15}$ $\frac{6}{10} = \frac{\_\_}{5}$

Name ______________________________ Date ______________

Simplest Form

A fraction is in **simplest form** when 1 is the only number that can divide both the numerator and the denominator evenly.

$\frac{4}{6}$ is *not* in simplest form because 4 and 6 are both divisible by 2. $\quad \frac{\mathbf{4} \div 2}{\mathbf{6} \div 2} = \frac{\mathbf{2}}{\mathbf{3}}$

$\frac{2}{3}$ is in simplest form because the only number that divides 2 and 3 evenly is 1. $\quad \frac{\mathbf{2} \div 1}{\mathbf{3} \div 1} = \frac{\mathbf{2}}{\mathbf{3}}$

Look at each row. Circle the fraction that is not in simplest form. Then write it in simplest form.

A $\frac{1}{4}$ $\quad \frac{2}{4}$ $\quad \frac{3}{4}$ ______

B $\frac{11}{12}$ $\quad \frac{7}{12}$ $\quad \frac{9}{12}$ ______

C $\frac{7}{8}$ $\quad \frac{5}{8}$ $\quad \frac{2}{8}$ ______

D $\frac{5}{10}$ $\quad \frac{3}{10}$ $\quad \frac{7}{10}$ ______

E $\frac{3}{6}$ $\quad \frac{1}{6}$ $\quad \frac{5}{6}$ ______

F $\frac{8}{9}$ $\quad \frac{6}{9}$ $\quad \frac{4}{9}$ ______

Write each fraction in simplest form.

G $\frac{2}{6} = \frac{\square}{\square}$ $\quad \frac{3}{9} = \frac{\square}{\square}$ $\quad \frac{8}{10} = \frac{\square}{\square}$ $\quad \frac{3}{12} = \frac{\square}{\square}$

H $\frac{4}{8} = \frac{\square}{\square}$ $\quad \frac{10}{12} = \frac{\square}{\square}$ $\quad \frac{3}{15} = \frac{\square}{\square}$ $\quad \frac{6}{8} = \frac{\square}{\square}$

I $\frac{4}{12} = \frac{\square}{\square}$ $\quad \frac{6}{10} = \frac{\square}{\square}$ $\quad \frac{6}{12} = \frac{\square}{\square}$ $\quad \frac{10}{15} = \frac{\square}{\square}$

Name ______________________________ Date ______________

Fractional Part of a Number

You can use pictures of equal groups to find the fraction of a number.

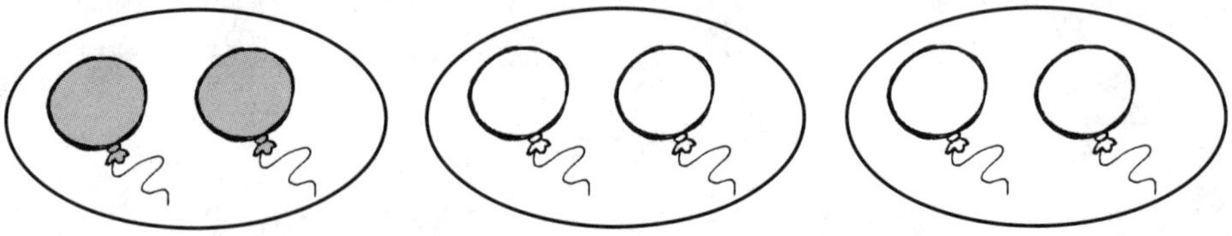

$\frac{1}{3}$ of 6 = **2**

Circle to make 2 equal groups. Then color the balloons in 1 group to help you solve each problem.

A

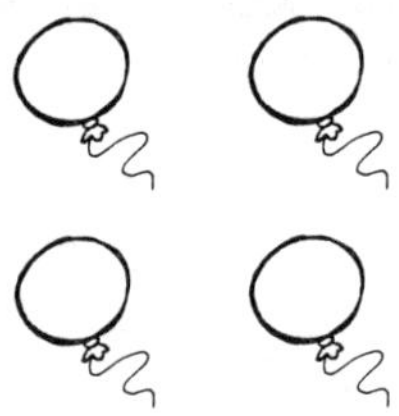

$\frac{1}{2}$ of 4 = ________

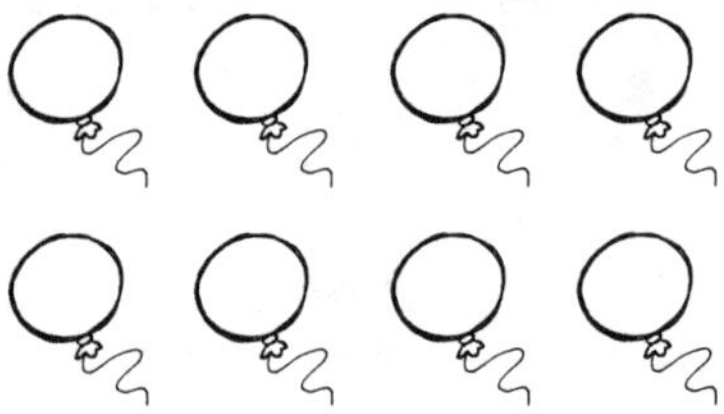

$\frac{1}{2}$ of 8 = ________

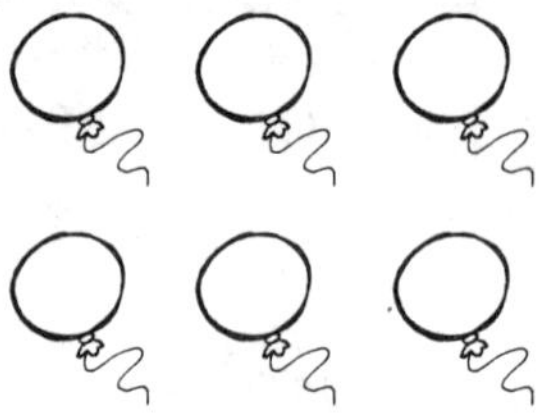

$\frac{1}{2}$ of 6 = ________

Circle to make 3 equal groups. Then color the balloons in 1 group to help you solve each problem.

B

$\frac{1}{3}$ of 3 = ________

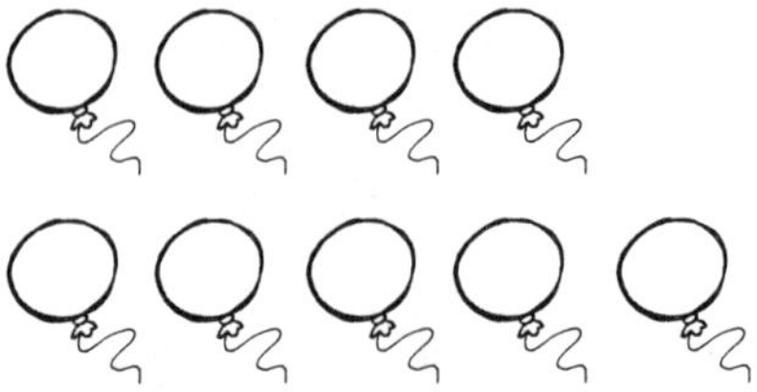

$\frac{1}{3}$ of 9 = ________

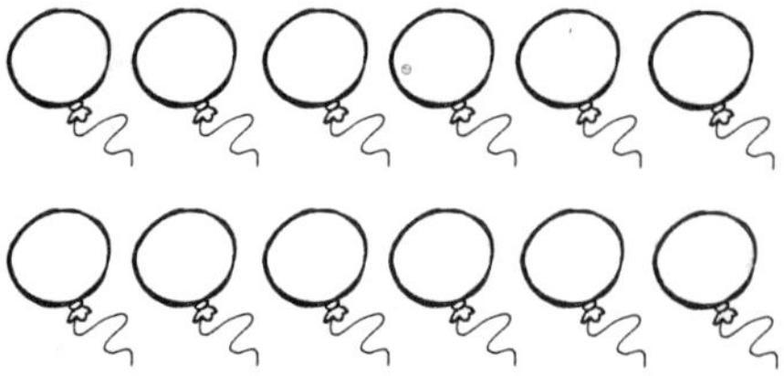

$\frac{1}{3}$ of 12 = ________

Circle to make 4 equal groups. Then color the balloons in 1 group to help you solve each problem.

C

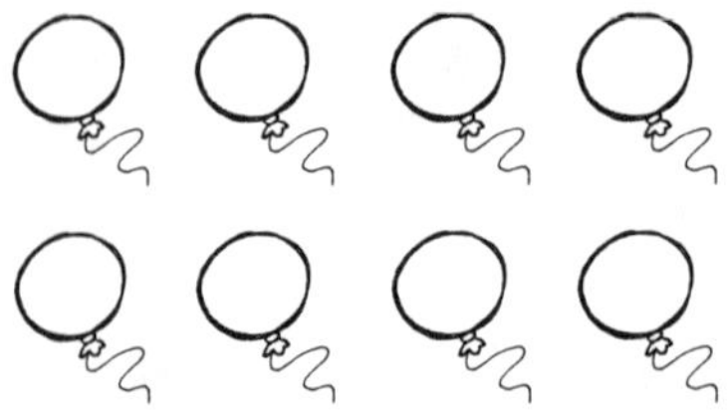

$\frac{1}{4}$ of 8 = ________

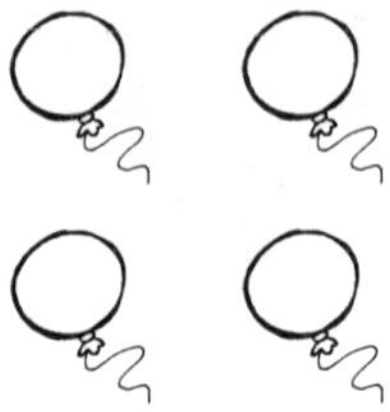

$\frac{1}{4}$ of 4 = ________

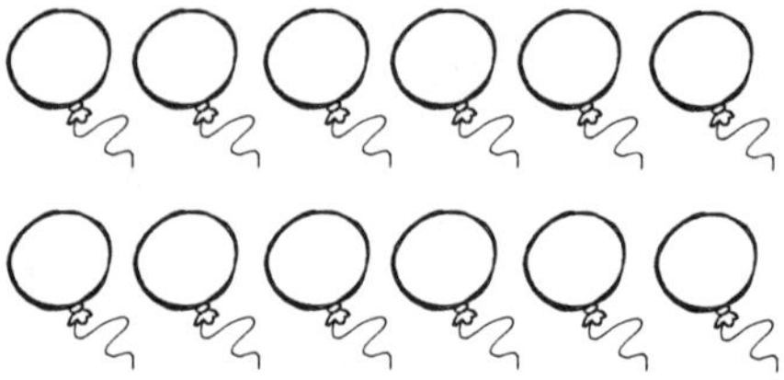

$\frac{1}{4}$ of 12 = ________

Name ______________________________ Date ____________

Parts of a Number

There are 20 beads.
Three fourths of them are blue.
What is ¾ of 20?

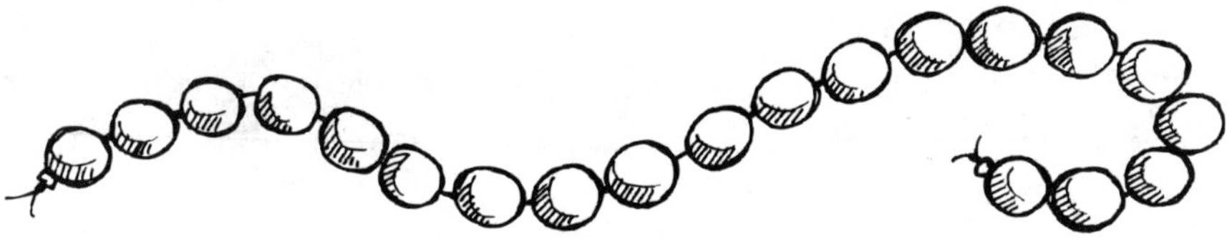

You can use division and multiplication to help you find the answer.

1. First, divide 20 by 4 to find out how many beads there are in 4 equal groups.

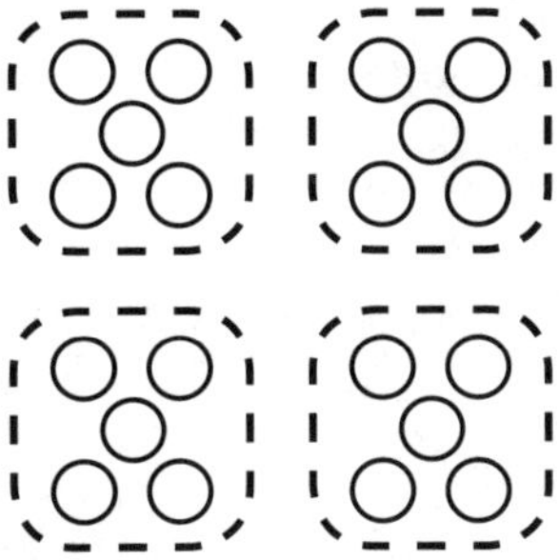

$20 \div 4 = 5$

There are 5 beads in each group.

2. Next, multiply that number by 3 to find out how many beads there are in 3 groups.

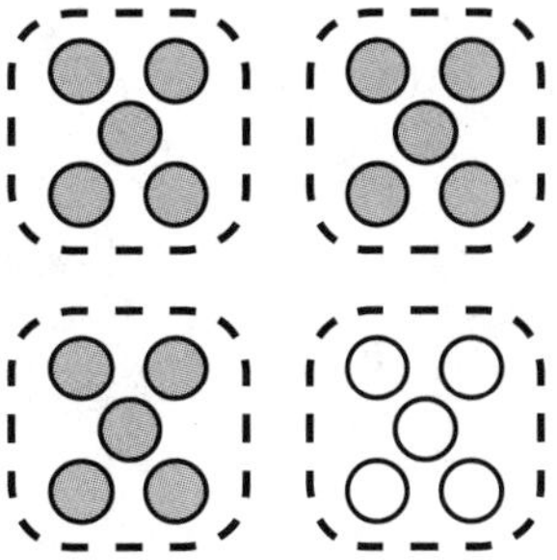

$3 \times 5 = 15$

There are 15 beads in 3 groups. That means ¾ of 20 equals 15.

Solve.

A $\frac{2}{3}$ of 15 = ______ $\frac{4}{6}$ of 12 = ______ $\frac{1}{8}$ of 16 = ______

B $\frac{3}{4}$ of 8 = ______ $\frac{2}{7}$ of 21 = ______ $\frac{2}{5}$ of 20 = ______

C $\frac{8}{9}$ of 18 = ______ $\frac{5}{8}$ of 16 = ______ $\frac{3}{9}$ of 18 = ______

D $\frac{5}{7}$ of 14 = ______ $\frac{3}{5}$ of 15 = ______ $\frac{2}{3}$ of 12 = ______

E $\frac{7}{10}$ of 20 = ______ $\frac{5}{6}$ of 18 = ______ $\frac{4}{5}$ of 15 = ______

Name ______________________________ Date ______________

Improper Fractions and Mixed Numbers

Matt cut some paper circles into halves. You can describe his shapes by using an **improper fraction** or a **mixed number**.

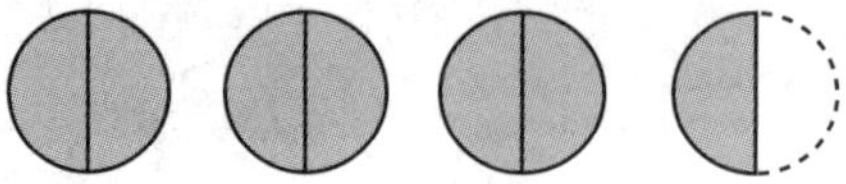

An improper fraction has a numerator that is greater than or equal to the denominator. A mixed number has a whole number and a fraction.

$$\frac{7}{2} = 3\frac{1}{2}$$

Write an improper fraction and a mixed number to describe the shaded parts in each picture.

A

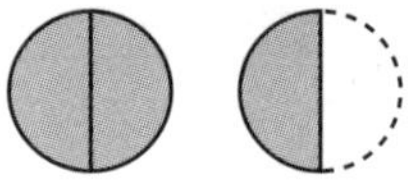

____

B

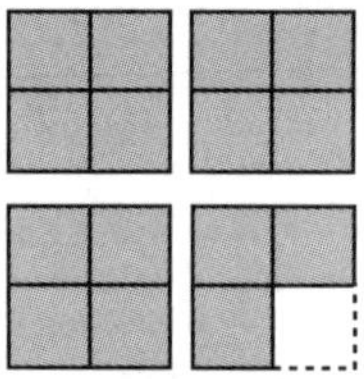

____

C

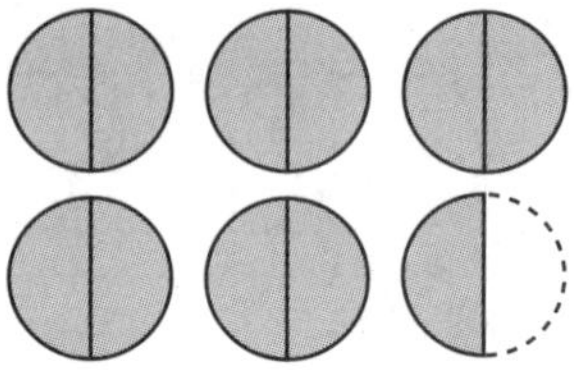

____

D

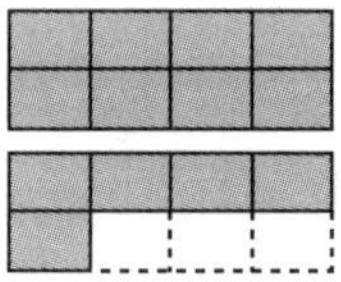

____

E

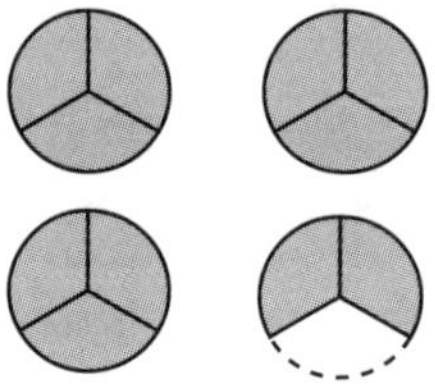

____

F

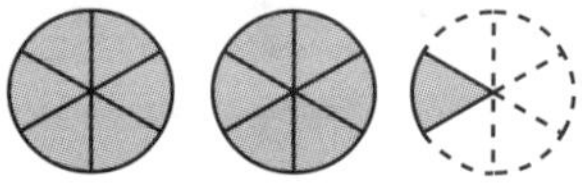

____

Some improper fractions equal a whole number. Write an improper fraction and a whole number to describe the shaded parts in each picture below.

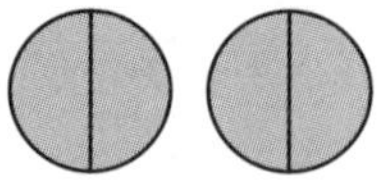

$$\frac{4}{2} = 2$$

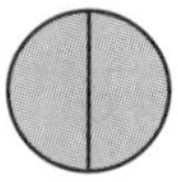

$$\frac{2}{2} = 1$$

G

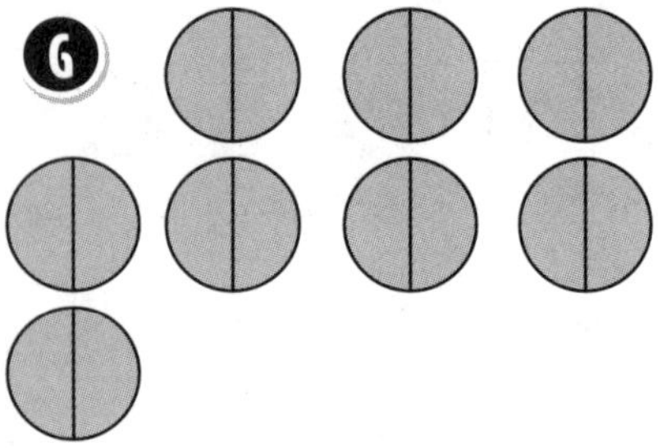

____

H

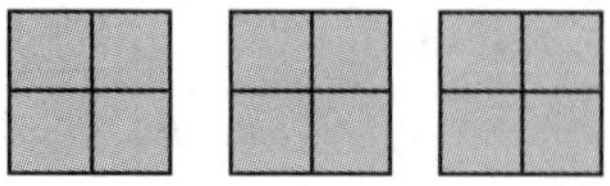

____

I

____

Name ______________________________ Date ______________

Changing Improper Fractions to Mixed Numbers

Megan has 5 sandwich halves. She groups the halves together to make whole sandwiches. How many whole sandwiches can she make? The picture shows that Megan will make 2 whole sandwiches and have 1 half left over.

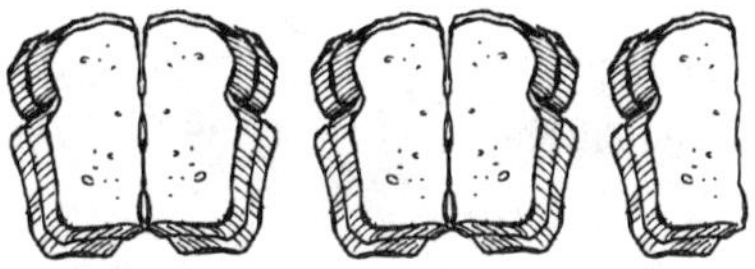

$$\frac{5}{2} = 2\frac{1}{2}$$

To change an improper fraction to a mixed number, divide the numerator by the denominator to get the number of wholes and the number of fractional parts.

$$\mathbf{\frac{5}{2}} \rightarrow \mathbf{2\overline{)5}}$$

2 ← number of wholes

$\quad 5$
-4
$\quad 1$ ← number of halves left over

Change the improper fractions to mixed numbers or whole numbers.

A $\frac{3}{2}$ = ________ $\frac{6}{3}$ = ________ $\frac{7}{5}$ = ________ $\frac{9}{4}$ = ________

B $\frac{9}{3}$ = ________ $\frac{11}{2}$ = ________ $\frac{15}{4}$ = ________ $\frac{8}{7}$ = ________

C $\frac{8}{4}$ = ________ $\frac{12}{5}$ = ________ $\frac{11}{8}$ = ________ $\frac{15}{3}$ = ________

D Kelly made sandwiches for her party. She cut the sandwiches into halves and placed them on a large plate. Kelly counted 16 sandwich halves. How many whole sandwiches did she make?

________ sandwiches

E Mike cut sandwiches into fourths. When he was done, he counted 24 fourths in all. How many whole sandwiches did Mike cut?

________ sandwiches

Name ______________________________ Date ______________

Changing Mixed Numbers to Improper Fractions

Betty's Bakery has only 2 ¼ pies left. How many fourths is that? To find out, first figure out how many fourths there are in 2 wholes. Do this by multiplying 2 × 4. That gives you 8 fourths. Then add that number to the remaining fourth to get 9 fourths in all.

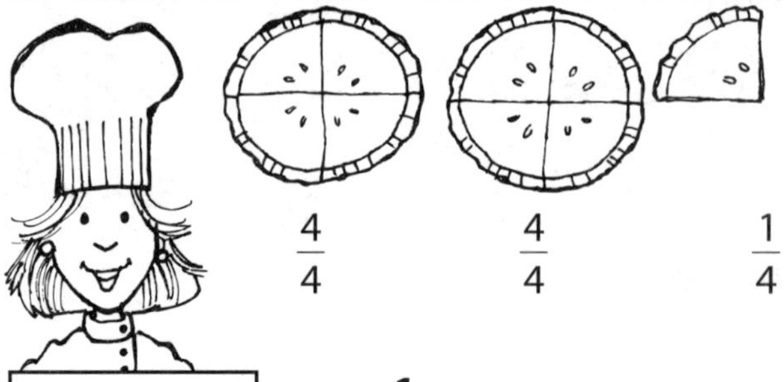

$\frac{4}{4}$ $\frac{4}{4}$ $\frac{1}{4}$

To change a mixed number to an improper fraction, first multiply the whole number by the denominator. This tells you the number of parts there are in the whole number. Then add that number to the number of leftover parts. This will give you the total number of parts.

$$2\frac{1}{4} = \frac{9}{4}$$

$$\mathbf{2\frac{1}{4} \longrightarrow (2 \times 4) + 1 = 9}$$

number of fourths in 2 → (2 × 4); number of leftover fourths → 1; total number of fourths → 9

Change the mixed numbers to improper fractions.

A $1\frac{1}{2}$ = ______ $1\frac{3}{4}$ = ______ $1\frac{5}{8}$ = ______ $1\frac{1}{3}$ = ______

B $2\frac{1}{2}$ = ______ $3\frac{1}{4}$ = ______ $5\frac{4}{5}$ = ______ $2\frac{3}{8}$ = ______

C $3\frac{2}{3}$ = ______ $6\frac{1}{10}$ = ______ $4\frac{1}{3}$ = ______ $3\frac{4}{5}$ = ______

D Glen brought 10 bagels to school. The teacher cut them into thirds. There are 21 students in Glen's class. Are there enough pieces for each student to get ⅓ bagel? Explain.

____

____

E Maria's cake recipe uses 3 ½ cups of flour. She only has a ½-cup measuring cup. How many times will Maria need to fill the cup?

________ times

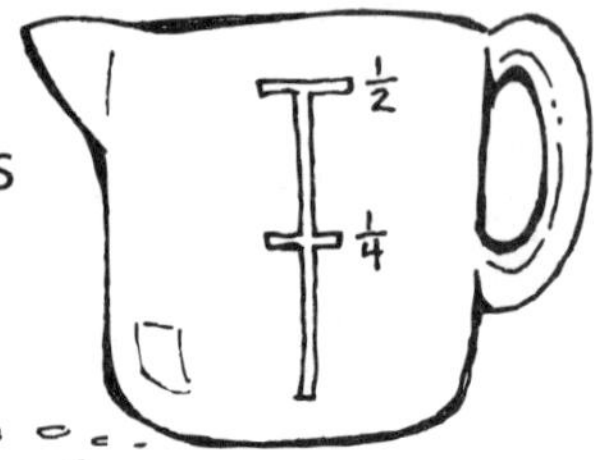

Name ______________________ Date ____________

Adding and Subtracting Fractions with Like Denominators

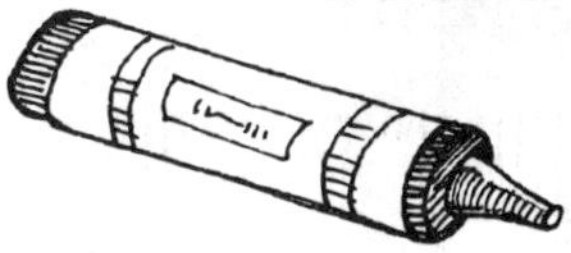

Color the parts to show addition. Then write the answers.

A

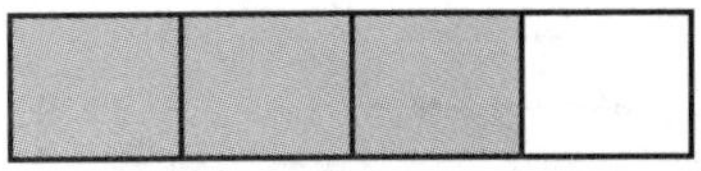

$\frac{1}{4}+\frac{2}{4}=\frac{3}{4}$

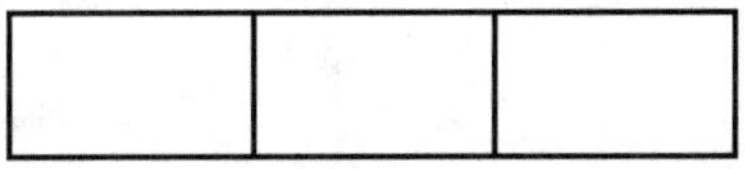

$\frac{1}{3}+\frac{1}{3}=$

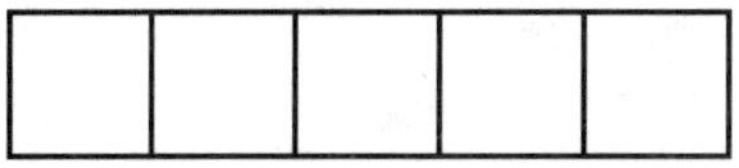

$\frac{2}{5}+\frac{2}{5}=$

B

$\frac{1}{6}+\frac{4}{6}=$

$\frac{1}{5}+\frac{2}{5}=$

$\frac{3}{8}+\frac{2}{8}=$

Cross out the parts to show subtraction. Then write the answers.

C

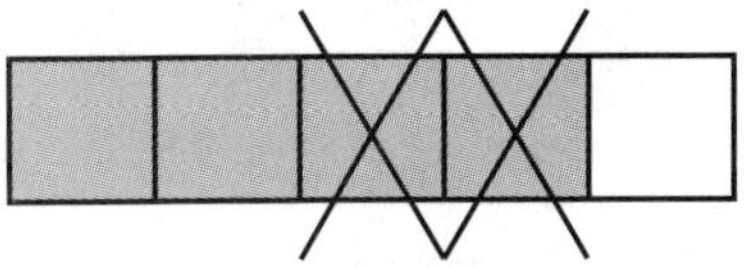

$\frac{4}{5}-\frac{2}{5}=\frac{2}{5}$

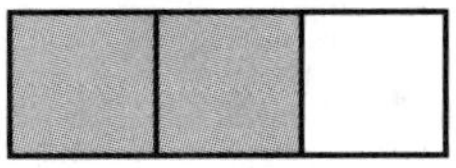

$\frac{2}{3}-\frac{1}{3}=$

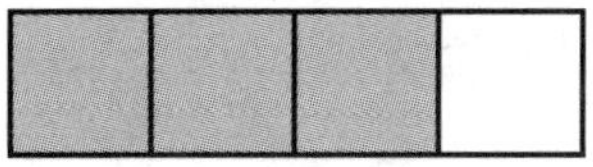

$\frac{3}{4}-\frac{1}{4}=$

D

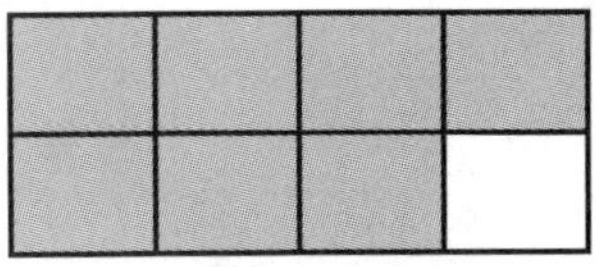

$\frac{7}{8}-\frac{4}{8}=$

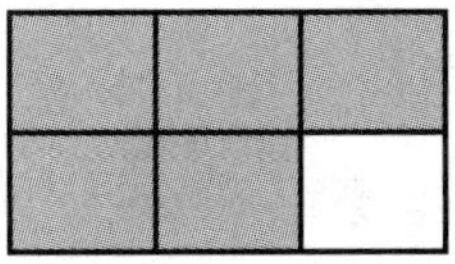

$\frac{5}{6}-\frac{4}{6}=$

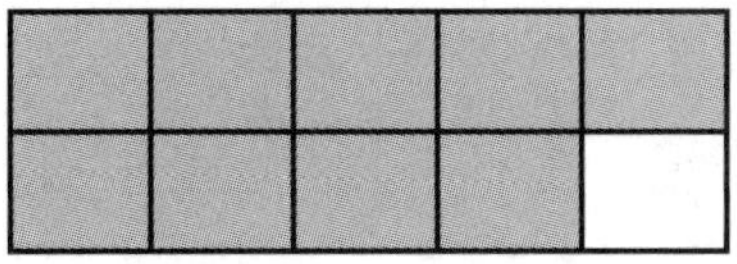

$\frac{9}{10}-\frac{7}{10}=$

Add or subtract.

E $\frac{2}{7}+\frac{3}{7}=$ $\frac{6}{9}-\frac{4}{9}=$ $\frac{3}{10}+\frac{4}{10}=$ $\frac{4}{6}-\frac{2}{6}=$

F $\frac{2}{9}+\frac{5}{9}=$ $\frac{7}{10}-\frac{6}{10}=$ $\frac{1}{7}+\frac{5}{7}=$ $\frac{6}{10}+\frac{3}{10}=$

Name ______________________________ Date ______________

Adding Fractions with Unlike Denominators

When adding two fractions that have different denominators, first change the fractions to equivalent fractions that have the same denominator. Then add.

Add.

A $\frac{1}{3}+\frac{3}{6}=$

B $\frac{2}{6}+\frac{1}{12}=$

C $\frac{2}{5}+\frac{5}{10}=$

D $\frac{3}{8}+\frac{1}{2}=$

E $\frac{5}{8}+\frac{1}{4}=$

F $\frac{2}{3}+\frac{2}{9}=$

G $\frac{1}{2}+\frac{1}{5}=$

H $\frac{1}{3}+\frac{1}{4}=$

I $\frac{2}{3}+\frac{1}{4}=$

J $\frac{3}{4}+\frac{3}{16}=$

K Mr. Morgan is driving his truck to Boomsville. By 11:00, he has driven ¾ of the distance. He drives another ⅛ of the way by noon. What fraction of the trip has he completed by noon? ________ of the trip

L There are several trucks rolling along the highway. Two fifths of the trucks are red and ¼ of them are black. What fraction of the trucks are either red or black?
________ of the trucks

Name ______________________________ Date ______________

Subtracting Fractions with Unlike Denominators

When subtracting fractions that have different denominators, first change the fractions to equivalent fractions that have the same denominator. Then subtract.

$$\frac{3}{4} - \frac{1}{8} = \frac{\mathbf{6}}{\mathbf{8}} - \frac{1}{8} = \frac{5}{8}$$

$$\frac{1}{2} - \frac{1}{3} = \frac{\mathbf{3}}{\mathbf{6}} - \frac{\mathbf{2}}{\mathbf{6}} = \frac{1}{6}$$

Subtract.

A $\frac{2}{3} - \frac{1}{9} =$

B $\frac{7}{12} - \frac{1}{6} =$

C $\frac{4}{6} - \frac{1}{2} =$

D $\frac{2}{5} - \frac{1}{4} =$

E $\frac{7}{8} - \frac{1}{2} =$

F $\frac{2}{3} - \frac{3}{7} =$

G $\frac{1}{3} - \frac{1}{4} =$

H $\frac{5}{8} - \frac{1}{2} =$

I $\frac{3}{5} - \frac{1}{4} =$

J $\frac{2}{3} - \frac{5}{9} =$

K Bert ran ⅔ of a mile on Friday. He ran ¼ of a mile on Saturday. How much farther did he run on Friday than on Saturday?

________ mile

L Marcy had 7⁄9 yard of ribbon. She used ⅓ yard for an art project. How much ribbon did she have left?

________ yard

Name ______________________________ Date ______________

Adding and Subtracting Fractions

Add or subtract. Write the answers in simplest form.

$\frac{5}{8}+\frac{1}{8}=\frac{6}{8}=\frac{6 \div 2}{8 \div 2}=\frac{3}{4}$	$\frac{11}{12}-\frac{3}{12}=\frac{8}{12}=\frac{8 \div 4}{12 \div 4}=\frac{2}{3}$

Add or subtract. Write answers in simplest form.

A $\frac{1}{6}+\frac{2}{6}=$

B $\frac{7}{15}+\frac{5}{15}=$

C $\frac{3}{8}-\frac{1}{8}=$

D $\frac{11}{12}-\frac{8}{12}=$

E $\frac{4}{12}+\frac{6}{12}=$

F $\frac{3}{4}-\frac{1}{4}=$

G $\frac{7}{9}-\frac{4}{9}=$

H $\frac{1}{12}+\frac{1}{12}=$

I $\frac{9}{10}-\frac{3}{10}=$

J $\frac{15}{16}-\frac{5}{16}=$

Add or subtract. (Remember to rename the fractions in each problem so that they have the same denominator.) Write the answers in simplest form.

K $\frac{1}{4}+\frac{1}{12}=$

L $\frac{1}{6}+\frac{7}{12}=$

M $\frac{8}{10}-\frac{3}{5}=$

N $\frac{5}{9}-\frac{1}{18}=$

O $\frac{2}{3}+\frac{3}{9}=$

P $\frac{7}{9}+\frac{1}{18}=$

Name ______________________________ Date ______________

Adding Mixed Numbers

When adding mixed numbers, add the fractions first. Change any improper fractions to mixed numbers. Then add the whole numbers. Write the answer in simplest form.

$$\begin{array}{r} 3\frac{1}{6} \\ +\,2\frac{1}{6} \\ \hline \end{array}$$

Add ⅙ and ⅙.

Add 3 and 2.

$$5\frac{2}{6} = 5\frac{2 \div 2}{6 \div 2} = 5\frac{1}{3}$$

$$\begin{array}{r} 2\frac{3}{5} \\ +\,5\frac{4}{5} \\ \hline \end{array}$$

Add ⅗ and ⅘.

Add 2 and 5.

$$7\frac{\mathbf{7}}{\mathbf{5}} = 7 + 1\frac{\mathbf{2}}{\mathbf{5}} = 8\frac{2}{5}$$

A

$$\begin{array}{r} 2\frac{1}{5} \\ +\,2\frac{2}{5} \\ \hline \end{array} \qquad \begin{array}{r} 5\frac{1}{6} \\ +\,4\frac{3}{6} \\ \hline \end{array} \qquad \begin{array}{r} 3\frac{3}{8} \\ +\,2\frac{4}{8} \\ \hline \end{array} \qquad \begin{array}{r} 6\frac{1}{2} \\ +\,4\frac{1}{2} \\ \hline \end{array}$$

B

$$\begin{array}{r} 1\frac{2}{3} \\ +\,4\frac{2}{3} \\ \hline \end{array} \qquad \begin{array}{r} 8\frac{3}{7} \\ +\,2\frac{1}{7} \\ \hline \end{array} \qquad \begin{array}{r} 6\frac{3}{10} \\ +\,5\frac{2}{10} \\ \hline \end{array} \qquad \begin{array}{r} 9\frac{3}{8} \\ +\,2\frac{6}{8} \\ \hline \end{array}$$

C

$$\begin{array}{r} 8\frac{1}{9} \\ +\,6\frac{2}{9} \\ \hline \end{array} \qquad \begin{array}{r} 4\frac{5}{12} \\ +\,3\frac{3}{12} \\ \hline \end{array} \qquad \begin{array}{r} 1\frac{8}{9} \\ +\,6\frac{6}{9} \\ \hline \end{array} \qquad \begin{array}{r} 7\frac{3}{10} \\ +\,8\frac{9}{10} \\ \hline \end{array}$$

Name ______________________ Date __________

Subtracting Mixed Numbers

When subtracting mixed numbers, subtract the fractions first. Then subtract the whole numbers. Write the answers in simplest form.

$$\begin{array}{r} 4\frac{7}{8} \\ -1\frac{3}{8} \\ \hline 3\frac{4}{8} \end{array}$$

Subtract ⅜ from ⅞.

Subtract 1 from 4.

$$3\frac{4}{8} = 3\frac{4 \div 4}{8 \div 4} = 3\frac{1}{2}$$

A

$$\begin{array}{r} 4\frac{8}{9} \\ -2\frac{4}{9} \\ \hline \end{array} \qquad \begin{array}{r} 6\frac{1}{2} \\ -5\frac{1}{2} \\ \hline \end{array} \qquad \begin{array}{r} 3\frac{5}{6} \\ -3\frac{3}{6} \\ \hline \end{array} \qquad \begin{array}{r} 7\frac{9}{10} \\ -4\frac{2}{10} \\ \hline \end{array}$$

B

$$\begin{array}{r} 5\frac{3}{4} \\ -2\frac{1}{4} \\ \hline \end{array} \qquad \begin{array}{r} 3\frac{4}{5} \\ -3\frac{2}{5} \\ \hline \end{array} \qquad \begin{array}{r} 7\frac{5}{8} \\ -6\frac{3}{8} \\ \hline \end{array} \qquad \begin{array}{r} 6\frac{11}{12} \\ -3\frac{10}{12} \\ \hline \end{array}$$

C

$$\begin{array}{r} 8\frac{9}{10} \\ -3\frac{5}{10} \\ \hline \end{array} \qquad \begin{array}{r} 4\frac{7}{9} \\ -1\frac{4}{9} \\ \hline \end{array} \qquad \begin{array}{r} 6\frac{7}{12} \\ -2\frac{5}{12} \\ \hline \end{array} \qquad \begin{array}{r} 9\frac{9}{10} \\ -3\frac{4}{10} \\ \hline \end{array}$$

Name ______________________________ Date ________________

Problem Solving with Fractions

Solve.

A Brian cut out 15 paper shapes. Two thirds of the shapes were circles. The rest were triangles. How many shapes were triangles?

__________ triangles

B Amy had 16 markers in a box. She put $\frac{5}{8}$ of the markers on her desk. How many markers were on her desk?

__________ markers

C There were 18 students working with clay. Two thirds of the class made pots. How many students made pots?

__________ students

D Cindy had yarn that was 20 inches long. She cut the yarn in half. Then she cut each piece in half again. How long was each piece?

__________ inches

E Mr. Jackson had 30 paintbrushes in a can. After he passed out the brushes to the class, only $\frac{1}{5}$ of the brushes were left in the can. How many students are in Mr. Jackson's class?

__________ students

F Mrs. Lane put her students' pictures on the wall. She arranged them in 6 rows with 4 pictures in each row. She saw that $\frac{1}{3}$ of the pictures were of pets. How many pictures showed pets?

__________ pictures

G Justin had a box of 12 crayons. He took one half of the crayons out of the box. Then he put 3 crayons back in the box. What fraction of the crayons were out of the box?

________ of the crayons

H Miss Wu's class made 36 paper flowers. One half of the flowers were red. One third of the flowers were yellow. The rest were blue. What fraction of the flowers were blue?

________ of the flowers

Name ______________________________ Date ______________

Fraction Review

Write the fraction that tells what part is shaded.

A 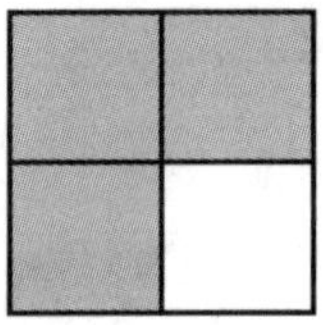 $\frac{\square}{\square}$ **B** 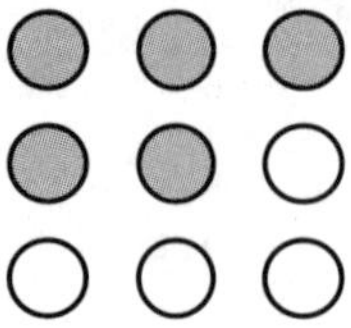 $\frac{\square}{\square}$

Compare the fractions. Write > or < in the circles.

C $\frac{4}{7} \bigcirc \frac{3}{7}$ $\frac{2}{6} \bigcirc \frac{2}{3}$ $\frac{5}{10} \bigcirc \frac{5}{8}$ $\frac{9}{12} \bigcirc \frac{4}{12}$

Write equivalent fractions.

D $\frac{1}{2} = \frac{\ }{6}$ $\frac{1}{4} = \frac{\ }{8}$ $\frac{1}{3} = \frac{\ }{6}$ $\frac{1}{2} = \frac{\ }{8}$

E $\frac{2}{3} = \frac{\ }{6}$ $\frac{2}{4} = \frac{\ }{2}$ $\frac{6}{8} = \frac{\ }{4}$ $\frac{8}{10} = \frac{\ }{5}$

Write each fraction in simplest form.

F $\frac{3}{9} = \text{—}$ $\frac{5}{10} = \text{—}$ $\frac{2}{8} = \text{—}$ $\frac{9}{12} = \text{—}$

Use the pictures to answer the questions.

G What is ½ of 6? ________

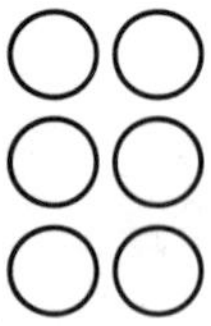

I What is ⅓ of 6? ________

H What is ¼ of 8? ________

J What is ¾ of 8? ________

Write an improper fraction and a mixed number to describe the shaded parts.

K ____________________ 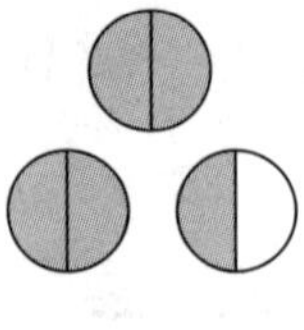

L ____________________ 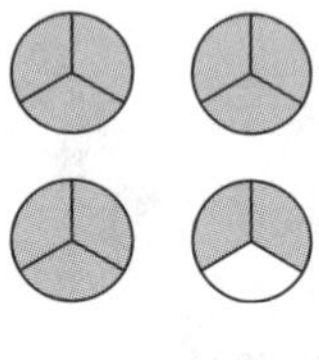

Name ______________________________ Date ______________

More Fraction Review

Add. Write the answers in simplest form.

A $\frac{1}{4} + \frac{1}{4} =$ $\frac{3}{6} + \frac{1}{6} =$ $\frac{2}{9} + \frac{4}{9} =$

B $\frac{2}{3} + \frac{2}{3} =$ $\frac{5}{8} + \frac{3}{8} =$ $\frac{9}{10} + \frac{9}{10} =$

C $\frac{1}{3} + \frac{1}{6} =$ $\frac{1}{2} + \frac{1}{4} =$ $\frac{3}{8} + \frac{1}{4} =$

D $2\frac{1}{5} + 4\frac{3}{5}$ $3\frac{1}{10} + 5\frac{1}{10}$ $5\frac{3}{8} + 1\frac{6}{8}$ $4\frac{2}{3} + 6\frac{1}{3}$

Subtract. Write the answers in simplest form.

E $\frac{5}{8} - \frac{2}{8} =$ $\frac{7}{10} - \frac{3}{10} =$ $\frac{8}{9} - \frac{2}{9} =$

F $\frac{1}{3} - \frac{1}{4} =$ $\frac{5}{8} - \frac{1}{2} =$ $\frac{11}{12} - \frac{2}{12} =$

Solve. Write your answer in simplest form.

G A pizza was divided into 8 equal pieces. Sarah ate ¼ of the pizza. Kyle ate twice as much as Sarah. What fraction of the pizza was left?

__________ of the pizza

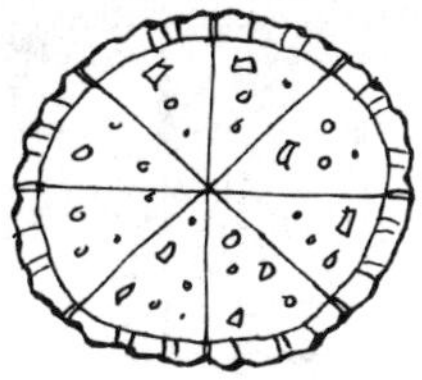

H Explain how you got the answer to problem G.

____

____

Name ______________________ Date ____________

Identifying Tenths

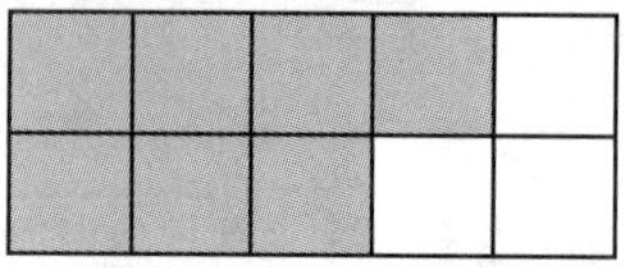

$\frac{7}{10}$ = 0.7 = seven tenths

Fill in the chart.

	Fraction	Decimal	Word Form
A 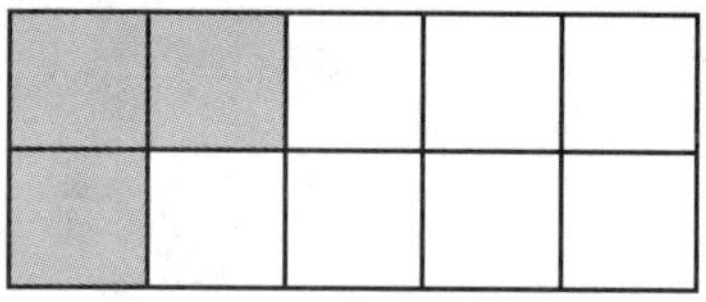	$\frac{3}{10}$	0.3	three tenths
B 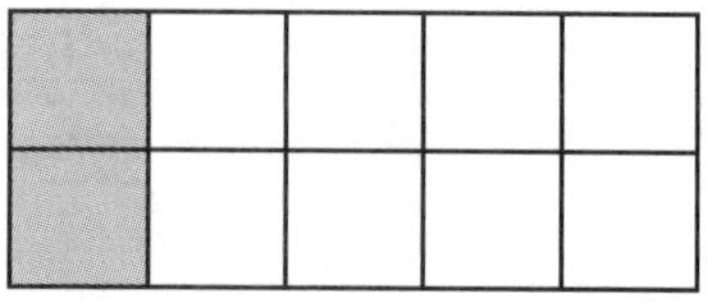			________ tenths
C 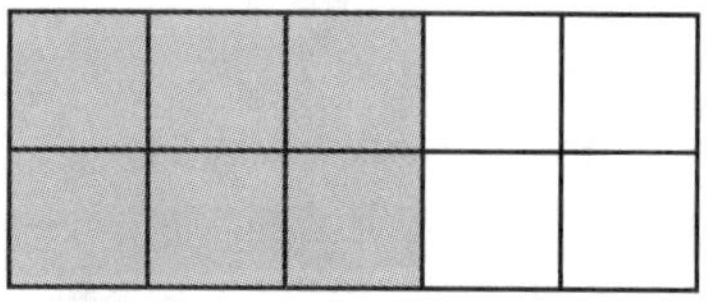			________ tenths
D 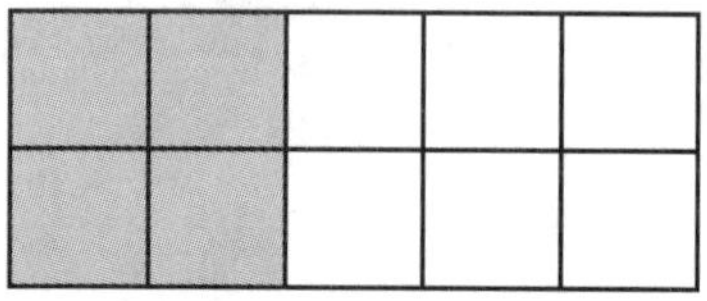			________ tenths
E 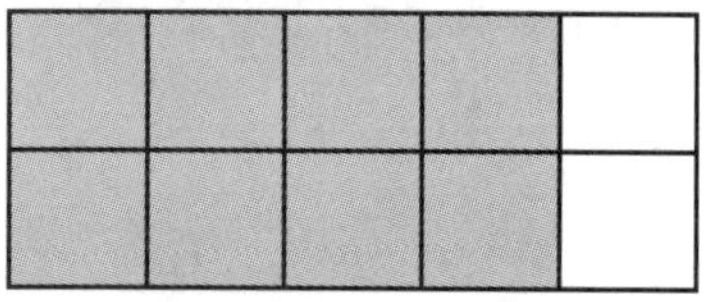			________ tenths
F 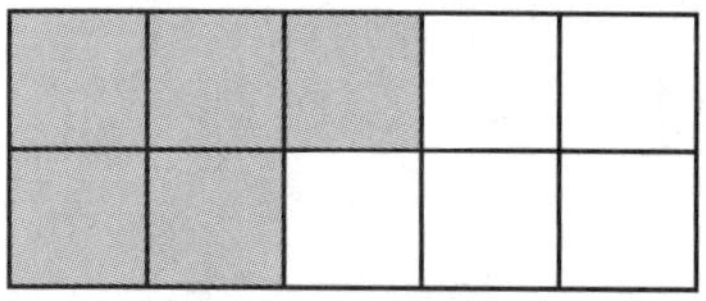			________ tenths
G 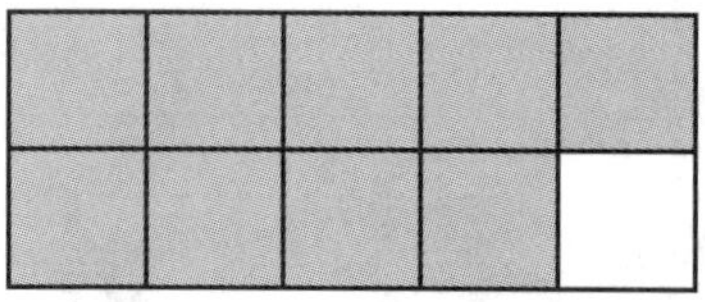			________ tenths

Name ______________________ Date ______________

Identifying Hundredths

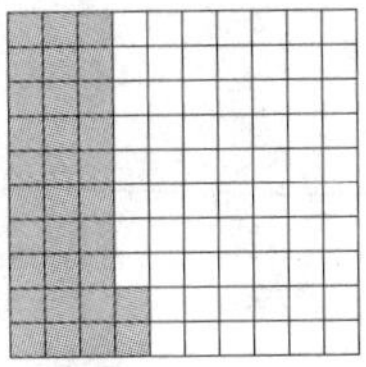

$\frac{32}{100} = 0.32$

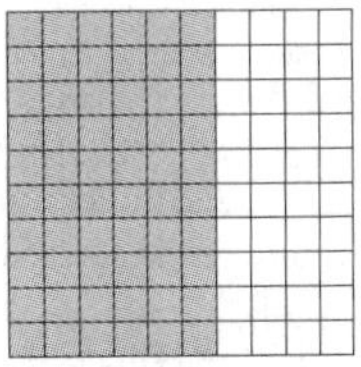

$\frac{60}{100} = 0.60$

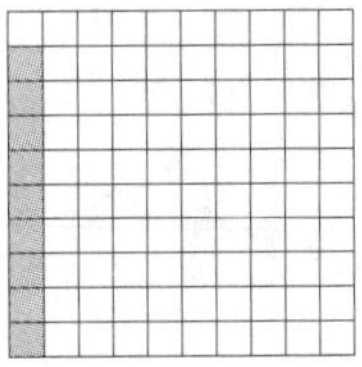

$\frac{9}{100} = 0.09$

Write a fraction and a decimal to describe the shaded part of each picture.

A

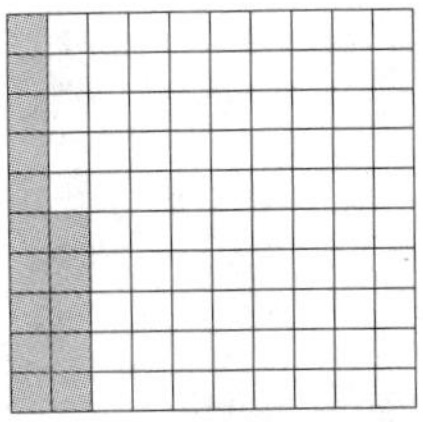

____

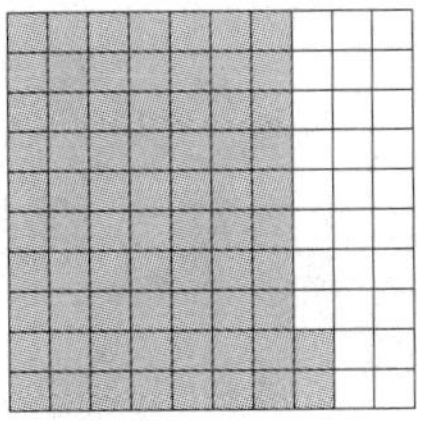

____

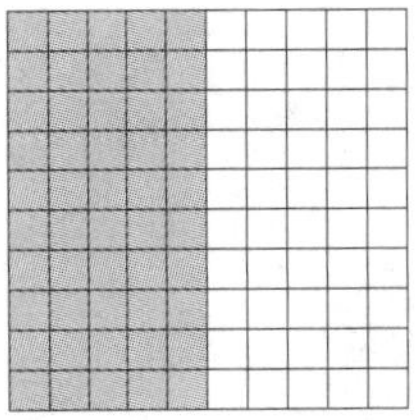

____

B

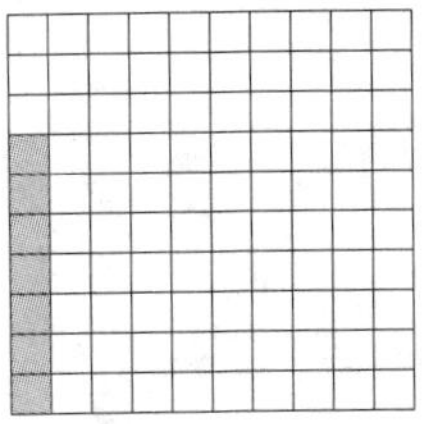

____

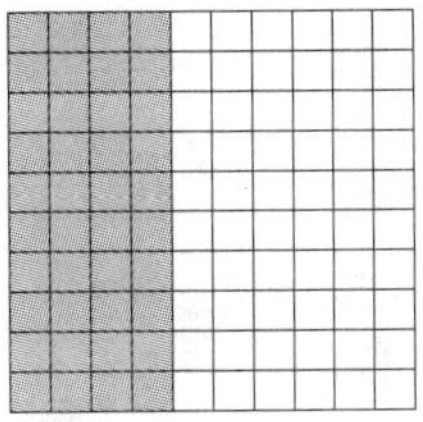

____

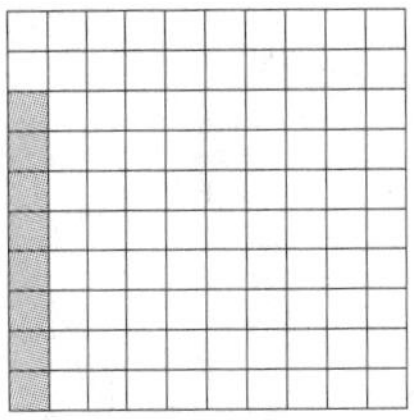

____

Color the squares to match each decimal.

C 0.55

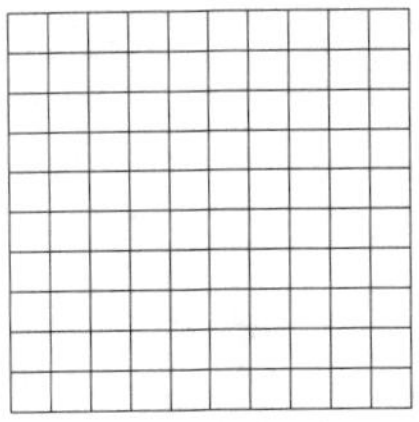

D 0.80

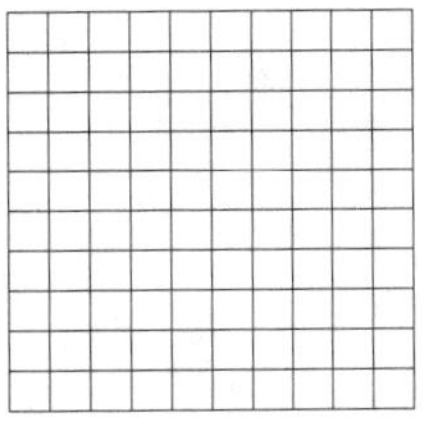

E 0.06

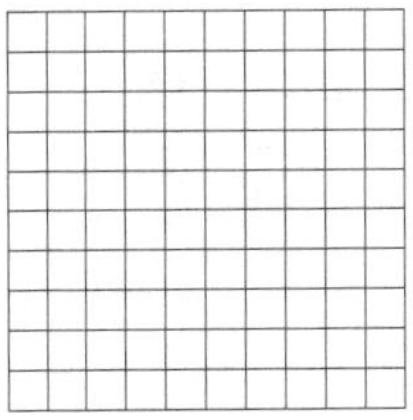

Name ______________________________ Date ______________

Decimals Greater Than 1

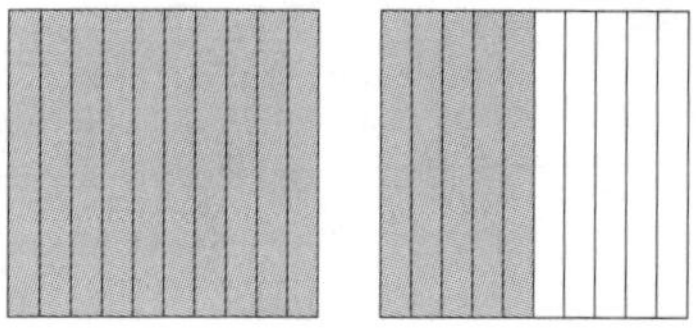

$1\frac{5}{10} = 1.5$

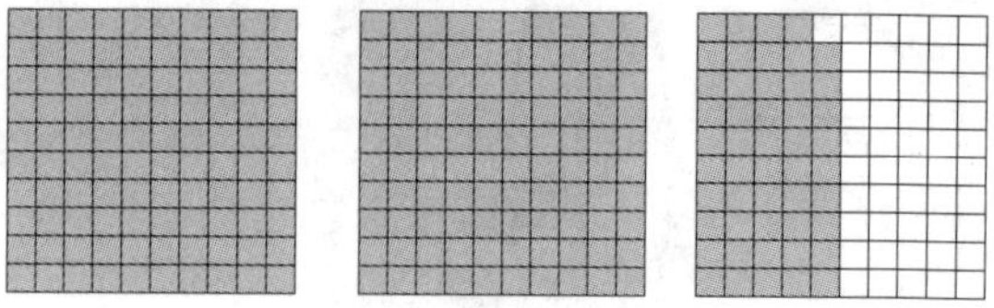

$2\frac{49}{100} = 2.49$

Write a mixed number and a decimal to describe the shaded parts.

A

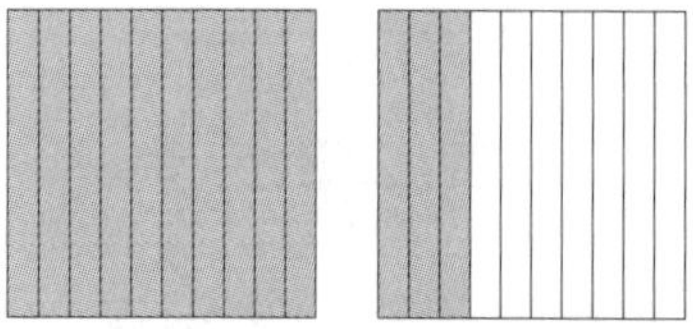

____

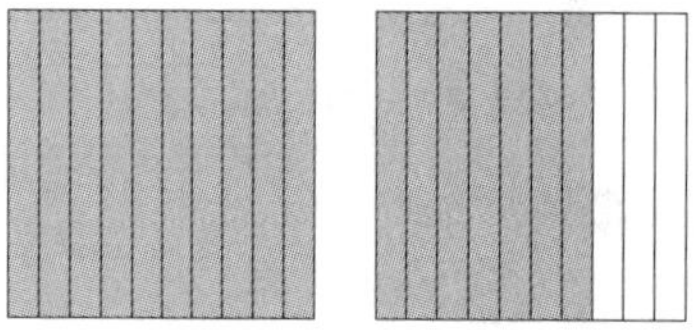

____

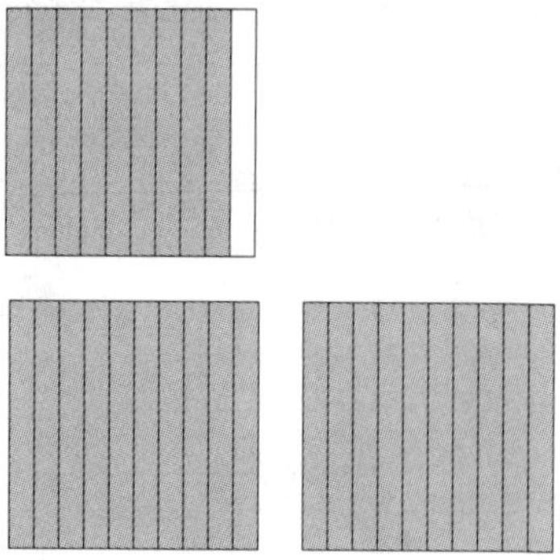

____

B

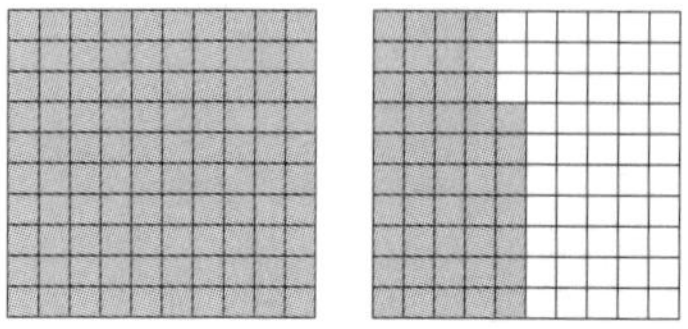

____

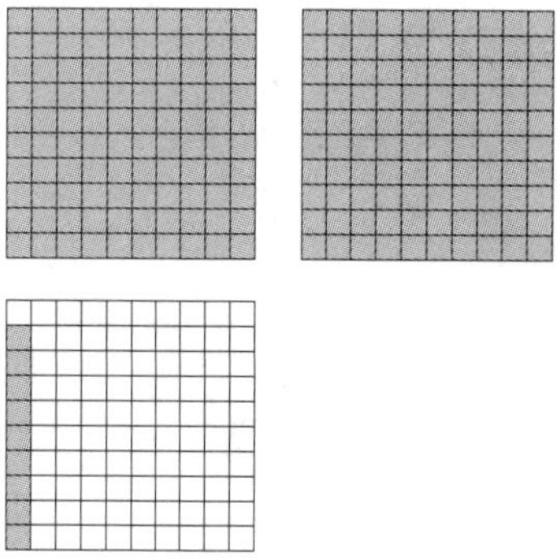

____

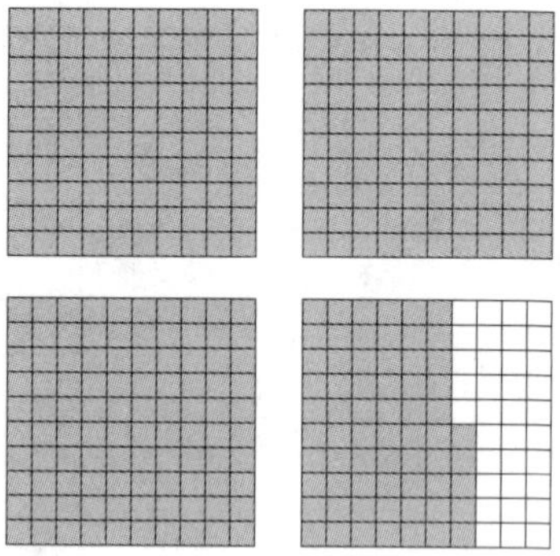

____

Write each mixed number as a decimal.

C $3\frac{16}{100}$ ________ $7\frac{5}{10}$ ________ $9\frac{89}{100}$ ________ $4\frac{1}{10}$ ________

Name ______________________ Date ____________

Matching Fractions and Decimals

Write the matching fractions.

A 0.3 ________ 0.4 ________ 0.9 ________ 0.17 ________

B 0.01 ________ 0.83 ________ 0.03 ________ 0.1 ________

Write the matching decimals.

C $\frac{2}{10}$ ________ $\frac{7}{10}$ ________ $\frac{95}{100}$ ________ $\frac{61}{100}$ ________

D $\frac{85}{100}$ ________ $\frac{2}{100}$ ________ $\frac{46}{100}$ ________ $\frac{5}{100}$ ________

Write the matching mixed numbers.

E 7.7 ________ 12.03 ________ 5.3 ________ 11.4 ________

F 8.09 ________ 4.59 ________ 6.02 ________ 15.01 ________

Write the matching decimals.

G $3\frac{1}{10}$ ________ $5\frac{7}{10}$ ________ $10\frac{35}{100}$ ________ $4\frac{65}{100}$ ________

H $9\frac{9}{100}$ ________ $14\frac{3}{10}$ ________ $17\frac{1}{10}$ ________ $7\frac{1}{100}$ ________

Name ______________________________ Date ______________

Comparing Tenths

Color the pictures to match the decimals. Then circle the greater decimal in each pair.

A 0.7

0.5

B 0.6

0.9

C 1.6

1.2

D 1.8

1.9

Compare the decimals. Write > or < in the circles.

E 0.2 ◯ 0.4   0.5 ◯ 0.1   0.9 ◯ 0.7   0.3 ◯ 0.4

F 1.5 ◯ 1.6   1.4 ◯ 1.6   2.3 ◯ 1.3   2.5 ◯ 5.2

Solve.

G Pat ran 1.2 miles. Kelly ran 1.3 miles. Who ran farther?

____

H Janet lives 2 blocks from school. Randy lives 1.9 blocks from school. Who lives farther from school?

____

Name ______________________________ Date ______________

Ordering Tenths

Which decimals are missing from the number line? Write them beside the matching letters. Use the decimals in the box.

0.5	1.2	0.3	0.8
0.9	1.6	0.2	1.3

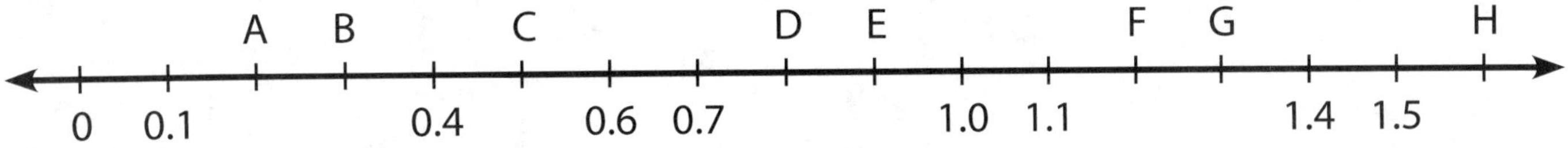

A ____________ **B** ____________ **C** ____________ **D** ____________

E ____________ **F** ____________ **G** ____________ **H** ____________

Look at the decimals in each set. Write them in order from least to greatest.

I

0.9	____________
0.2	____________
0.3	____________
0.7	____________

J

1.3	____________
1.6	____________
1.8	____________
1.5	____________

K

2.0	____________
1.3	____________
0.6	____________
0.9	____________

Write the next four decimals.

L 0.1, 0.2, 0.3, 0.4, ________, ________, ________, ________

M 5.1, 5.2, 5.3, 5.4, ________, ________, ________, ________

N 7.5, 7.6, 7.7, 7.8, ________, ________, ________, ________

O 4.9, 4.8, 4.7, 4.6, ________, ________, ________, ________

P 8.6, 8.5, 8.4, 8.3, ________, ________, ________, ________

Q 12.4, 12.3, 12.2, 12.1, ________, ________, ________, ________

Name ______________________________ Date ______________

Comparing and Ordering Hundredths

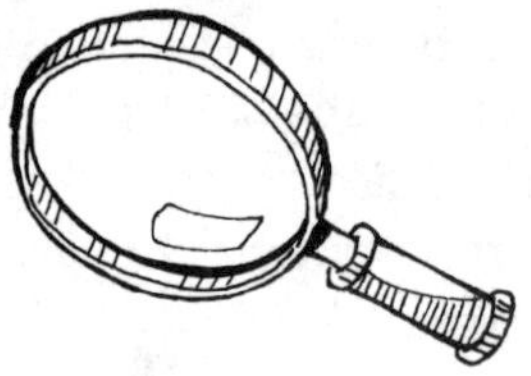

Color the pictures to match the decimals. Then compare the decimals in each pair. Write >, <, or = in the circles.

A

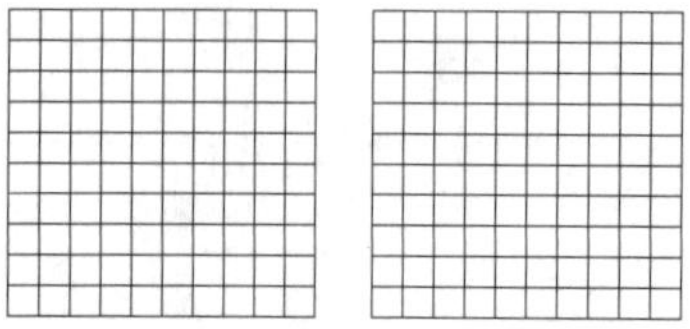

0.75 ◯ 0.57

B

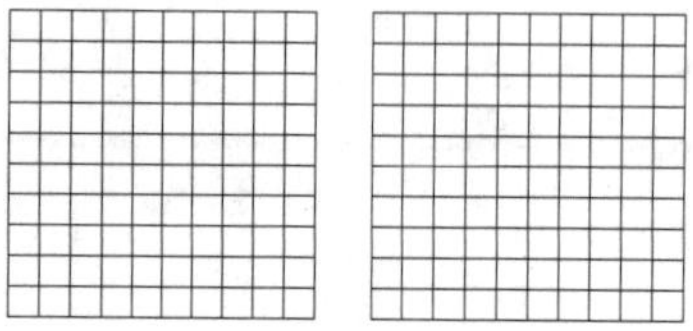

0.90 ◯ 0.98

C

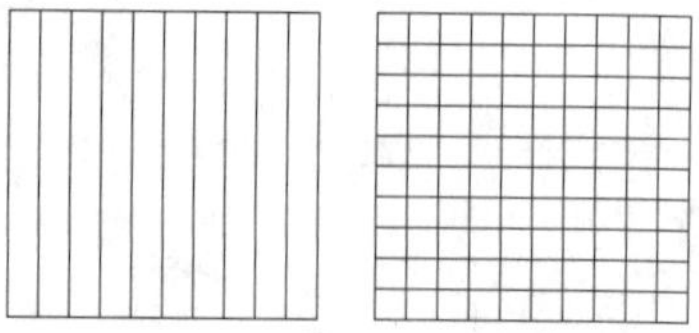

0.4 ◯ 0.40

D

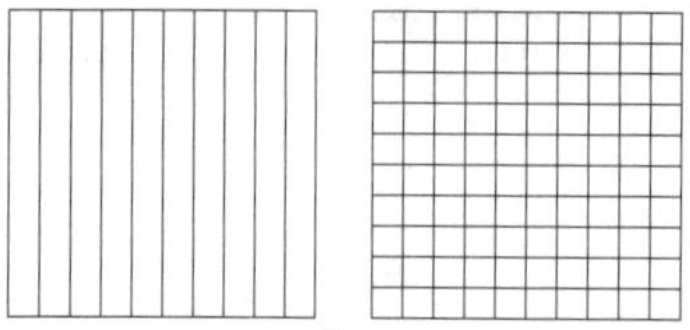

0.1 ◯ 0.16

E

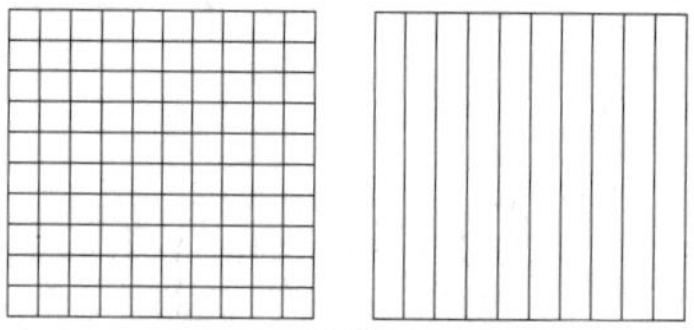

0.30 ◯ 0.3

F

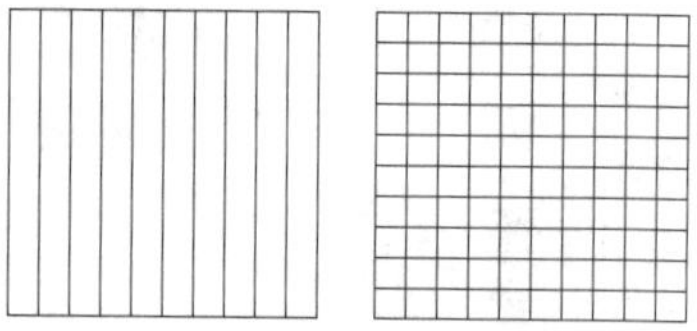

0.5 ◯ 0.29

When comparing tenths and hundredths, try changing the tenths to hundredths first. Do this by adding 0 to the right of the tenth. Then compare.

Example: When comparing 3.4 and 3.17, first change 3.4 to 3.40. Since 3 and 40 hundredths is greater than 3 and 17 hundredths, 3.4 is greater than 3.17.

Compare the decimals. Write >, <, or = in the circles.

G 2.15 ◯ 2.2   1.7 ◯ 1.70   0.8 ◯ 0.91   3.6 ◯ 3.18

H 7.42 ◯ 7.24   9.06 ◯ 9.5   5.13 ◯ 5.31   8.2 ◯ 8.02

I 0.15 ◯ 0.1   7.04 ◯ 7.4   9.6 ◯ 9.60   4.09 ◯ 4.9

Name ______________________________ Date ______________

Comparing and Ordering Fractions and Decimals

Circle the greatest number. Draw a box around the smallest number.

A 0.23 0.32 $\frac{30}{100}$

B 2.1 2.15 $2\frac{5}{10}$

C 4.6 4.46 $4\frac{40}{100}$

D 9.78 9.8 $9\frac{7}{10}$

Write the numbers in order from least to greatest.

E

0.7	________
0.09	________
0.9	________
0.79	________

F

1.2	________
1.08	________
1.5	________
1.28	________

G

$\frac{4}{10}$	________
0.39	________
$\frac{16}{100}$	________
0.3	________

Find the decimals that are missing from the number line. Write them beside the matching letters. Use the numbers in the box.

1.4	1.82	0.35	0.75
1.25	0.9	0.2	1.7

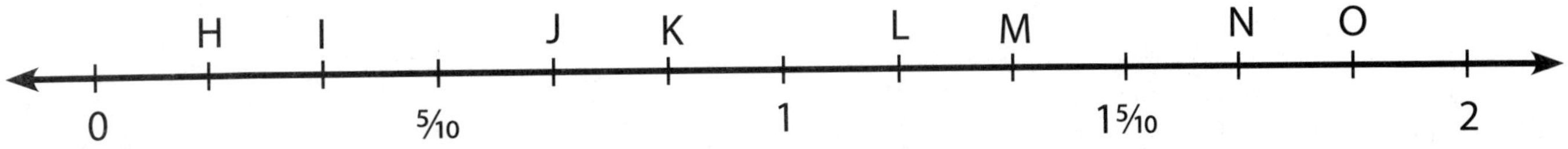

H ________ **I** ________ **J** ________ **K** ________

L ________ **M** ________ **N** ________ **O** ________

Name ______________________________ Date ______________

Rounding Decimals

When rounding decimals to the nearest whole number, look at the number in the **tenths** place. If it is 5 or greater, round up to the next whole number. If it is less than 5, keep the whole number as is.

When rounding decimals to the nearest tenth, look at the number in the **hundredths** place. If it is 5 or greater, round up to the next tenth. If it is less than 5, keep the tenth as is.

3.7 → The number in the tenths place is 7. Round 3.7 to 4.

3.2 → The number in the tenths place is 2. Round 3.2 to 3.

2.48 → The number in the hundredths place is 8. Round 2.48 to 2.5.

2.43 → The number in the hundredths place is 3. Round 2.43 to 2.4.

Round to the nearest whole number.

A 7.6 ________ 3.5 ________ 9.2 ________ 8.1 ________

B 3.4 ________ 9.6 ________ 12.8 ________ 11.4 ________

C 5.91 ________ 16.34 ________ 14.52 ________ 2.78 ________

Round to the nearest tenth.

D 1.38 ________ 6.07 ________ 5.82 ________ 9.49 ________

E 20.87 ________ 15.38 ________ 26.55 ________ 11.13 ________

F 0.63 ________ 17.03 ________ 40.09 ________ 32.75 ________

Solve.

G A bull stands 5.3 feet tall. What is its height rounded to the nearest foot?

________ feet

H A rope is 6.27 meters long. What is its length rounded to the nearest tenth of a meter?

________ meters

Name ______________________ Date ____________

Adding Tenths

Adding tenths is similar to adding whole numbers. Add the columns one at a time, beginning with the far right. Regroup if needed. Remember to bring down the decimal and include it in the answer.

0.5 + 2.3	1. Add the tenths. 2. Add the ones.
	0.5 + 2.3 = 2.8

4.6 + 3.7	1. Add the tenths. There are 13 tenths. Write down the 3 and place the 1 above the ones column.	2. Add the ones.
	(1) 4.6 + 3.7 = .3	(1) 4.6 + 3.7 = 8.3

Add.

A
$0.2 + 0.4$
$0.8 + 0.9$
$0.4 + 3.4$
$1.2 + 1.7$
$5.6 + 9.4$

B
$8.3 + 0.9$
$12.4 + 16.7$
$25.3 + 25.7$
$17.8 + 19.0$
$32.5 + 6.7$

C
$23.6 + 6.8$
$14.5 + 19.3$
$37.9 + 30.7$
$4.4 + 21.0$
$20.7 + 19.3$

Complete each chart.

D

Add 0.3	
Input	**Output**
5.0	
0.6	
1.9	

E

Add 1.5	
Input	**Output**
0.4	
1.5	
3.2	

F

Add 4.7	
Input	**Output**
2.1	
3.2	
5.9	

Name ______________________ Date __________

Subtracting Tenths

Subtracting tenths is similar to subtracting whole numbers. Subtract the columns one at a time beginning at the far right. Regroup if needed. Remember to bring down the decimal and include it in the answer.

$$\begin{array}{r} 5.9 \\ -\,1.2 \\ \hline \end{array}$$

1. Subtract the tenths.
2. Subtract the ones.

$$\begin{array}{r} 5.9 \\ -\,1.2 \\ \hline 4.7 \end{array}$$

$$\begin{array}{r} 8.5 \\ -\,3.6 \\ \hline \end{array}$$

1. Regroup the ones to make 7 ones and 10 tenths. Combine the 10 tenths with the 5 tenths to get 15 tenths.

$$\begin{array}{r} {}^{7}\ {}^{15} \\ \not{8}.\not{5} \\ -\,3.6 \\ \hline \end{array}$$

2. Subtract the tenths.

Then subtract the ones.

$$\begin{array}{r} {}^{7}\ {}^{15} \\ \not{8}.\not{5} \\ -\,3.6 \\ \hline 4.9 \end{array}$$

Subtract.

A	7.5 − 2.4	3.6 − 0.4	11.8 − 9.4	6.5 − 2.8	24.1 − 7.6
B	20.5 − 3.8	18.6 − 9.2	32.4 − 2.5	45.9 − 28.7	68.8 − 18.9
C	38.7 − 16.4	24.5 − 9.8	37.6 − 19.5	73.6 − 70.8	52.4 − 17.6

Solve.

D Mr. Cooper had a box of nails that weighed 18.5 ounces. After he used some nails, the box weighed 12.8 ounces. What was the weight of the nails that he used?

__________ ounces

E Becky's board is 60.3 centimeters. Matt's board is 49.5 centimeters. How much longer is Becky's board than Matt's?

__________ centimeters

Name ______________________________ Date ______________

Adding and Subtracting Tenths

Add or subtract.

A	4.3 + 2.5	0.9 + 6.9	7.6 + 0.3	15.2 + 3.8	28.1 + 7.6
B	12.3 − 1.3	18.4 − 3.6	20.9 − 9.4	31.4 − 12.5	25.4 − 6.3
C	36.0 + 0.8	25.7 + 16.4	59.3 − 13.8	28.4 − 6.8	15.2 + 15.8
D	54.6 − 48.3	62.9 − 17.8	43.2 + 9.9	7.3 + 20.8	66.3 − 23.6
E	2.6 3.5 + 4.2	1.2 13.4 + 9.5	0.6 0.5 + 7.4	9.3 18.2 + 14.5	3.8 6.4 + 12.3

Solve.

F Choose an addition problem you solved. Write a word problem that matches it.

____

____

____

G Choose a subtraction problem you solved. Write a word problem that matches it.

____

____

____

Name ______________________________ Date ______________

Adding Hundredths

Adding hundredths is similar to adding tenths. When adding tenths to hundredths, change the tenths to hundredths by adding a 0 in the hundredths place. Then add. Regroup if needed.

4.35
+ 2.13

1. Add the hundredths.
2. Add the tenths.
3. Add the ones.

$$\begin{array}{r} 4.35 \\ +\,2.13 \\ \hline 6.48 \end{array}$$

6.5
+ 3.25

1. Add 0 to the end of 6.5 so that it has two decimal places like 3.25.

$$\begin{array}{r} 6.5\mathbf{0} \\ +\,3.25 \\ \hline \end{array}$$

2. Add the columns one at a time, beginning with the hundredths.

$$\begin{array}{r} 6.50 \\ +\,3.25 \\ \hline 9.75 \end{array}$$

Add.

A	7.21 + 1.63	8.34 + 1.92	5.5 + 0.67	2.9 + 1.84	0.93 + 0.6
B	4.81 + 7.99	2.67 + 0.14	4.8 + 3.44	3.9 + 5.72	5.6 + 11.32
C	19.8 + 18.37	23.65 + 5.6	17.08 + 3.5	2.46 + 18.08	32.7 + 9.03

Solve.

D A can of oranges weighed 11.5 ounces. A can of pears weighed 15.75 ounces. How much did the cans weigh altogether?

__________ ounces

E One package of onion soup mix weighed 28.35 grams. What did two packages of the mix weigh?

__________ grams

Name ______________________ Date ____________

Subtracting Hundredths

Subtracting hundredths is similar to subtracting tenths. When subtracting tenths from hundredths, change the tenths to hundredths by adding a 0 in the hundredths place. Then subtract. Regroup if needed.

$$\begin{array}{r} 5.75 \\ -\ 4.23 \\ \hline \end{array}$$

1. Subtract the hundredths.
2. Subtract the tenths.
3. Subtract the ones.

$$\begin{array}{r} 5.75 \\ -\ 4.23 \\ \hline 1.52 \end{array}$$

$$\begin{array}{r} 6.4 \\ -\ 3.29 \\ \hline \end{array}$$

1. Add 0 to the end of 6.4 so that it has two decimal places like 3.29.

$$\begin{array}{r} 6.4\mathbf{0} \\ -\ 3.29 \\ \hline \end{array}$$

2. Subtract the columns one at a time, beginning with the hundredths.

$$\begin{array}{r} {\scriptstyle 3\ 10} \\ 6.\not{4}\not{0} \\ -\ 3.29 \\ \hline 3.11 \end{array}$$

Subtract.

A	9.86 − 5.34	2.34 − 0.13	14.85 − 6.49	26.85 − 9.8	37.4 − 16.24
B	45.32 − 19.1	57.86 − 34.3	23.92 − 0.8	15.73 − 6.99	65.27 − 3.1
C	18.07 − 3.5	39.1 − 16.55	40.83 − 29.2	63.5 − 28.46	50.07 − 9.5

Solve.

D Brent lives 10.75 miles from the beach. Jessica lives 4.8 miles from the beach. How much farther does Brent live from the beach than Jessica?

__________ miles

E Lori's surfboard is 6.25 feet long. Kent's surfboard is 7.4 feet long. How much longer is Kent's board than Lori's?

__________ feet

Name ______________________________ Date ______________

Problem Solving with Decimals

The chart shows the weights of five dogs. Use the chart to answer the questions.

Dog Weights	
Dog	**Weight in Pounds**
Ruff	25.45
Pepper	12.5
Wags	30.6
Scooter	10.39
Fangs	42.8

A How much do the two lightest dogs weigh altogether?

____

B How much do the two heaviest dogs weigh altogether?

____

C What is the difference in weight between the heaviest dog and the lightest dog?

____

D Which two dogs have a total weight of 35.84 pounds? ______________________

E Which dog weighs about 17 pounds less than Fangs? ______________________

F Is Scooter more or less than 10 ½ pounds? ______________________

How do you know? ____

____

G Suppose Wags and Pepper got on a scale together. Would they weigh more or less than Fangs?

____

How much would they weigh? ______________________

Name ______________________ Date ______________

Decimals, Fractions, and Money

One hundred pennies is worth 100 cents, or 1 dollar. You can write the amount as 100¢ or $1.00. You can use fractions to show how the value of coins can be written as a decimal amount.

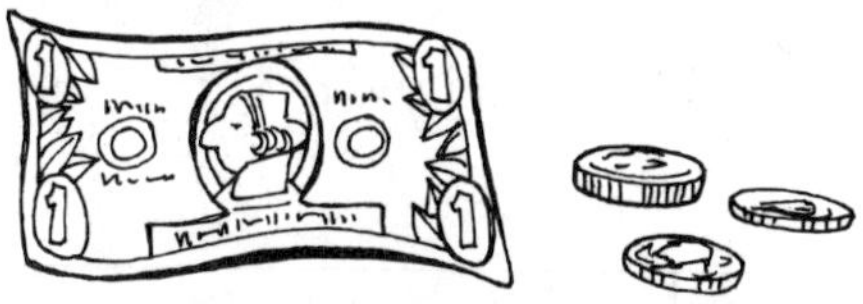

Fill in the chart.

Coin	Number of Cents	Fraction of a Dollar	Decimal Amount
penny	1¢	$\frac{1}{100}$	$0.01
nickel	______¢		
dime	______¢		
quarter	______¢		
half dollar	______¢		

Coin	Number of Cents	Fraction of a Dollar	Decimal Amount
8 pennies	______¢		
5 nickels	______¢		
7 dimes	______¢		
3 quarters	______¢		
2 half dollars	______¢		

Write each amount as a fraction of a dollar.

A 83¢ = __________ of a dollar

B $0.49 = __________ of a dollar

C 19¢ = __________ of a dollar

D $0.37 = __________ of a dollar

E 21¢ = __________ of a dollar

F $0.99 = __________ of a dollar

G Julie has 70 pennies. She says she has $\frac{7}{10}$ of a dollar. Is she right? Explain why or why not.

____

____

Name ______________________________ Date ______________

Decimal Patterns

Study the patterns. Then add the next four numbers to each pattern.

A 0.2, 0.4, 0.6, 0.8, ________, ________, ________, ________

What number was added each time? ________

B 0.05, 0.10, 0.15, 0.20, ________, ________, ________, ________

What number was added each time? ________

C 1.3, 1.2, 1.1, 1.0, ________, ________, ________, ________

What number was subtracted each time? ________

D 1.0, 0.89, 0.78, 0.67, ________, ________, ________, ________

What number was subtracted each time? ________

Use patterns to solve the problems.

E Lisa walked 0.5 miles on Monday. Each day after that, she walked 0.2 miles more than the day before. How many miles did she walk altogether after 4 days?

____

F Jake's plant grows 0.15 centimeters a day. On Sunday, his plant is 10.15 centimeters tall. How tall will the plant be next Friday?

____

G Carla arranged 5 bags of fruit on a shelf. Each bag was 0.25 pounds less than the bag to its left. The bag on the far left weighed 3.25 pounds. How much did the bag on the far right weigh?

____

Name ______________________________ Date ______________

Fraction and Decimal Equivalents

You can easily change certain fractions into decimals.

Fractions that have a denominator of 10 or 100
Use the numerator to help you write the decimal.

$\frac{3}{10} \rightarrow 0.3$

$\frac{45}{100} \rightarrow 0.45$

Fractions that can be changed to tenths or hundredths
Find an equivalent fraction that has a denominator of 10 or 100. Then use the numerator of that fraction to help you write the decimal.

$\frac{1}{2} \rightarrow \frac{1 \times 5}{2 \times 5} = \frac{5}{10} \rightarrow 0.5$

$\frac{1}{2} \rightarrow \frac{1 \times 50}{2 \times 50} = \frac{50}{100} \rightarrow 0.50$

Change the fractions to decimals. If you need to, change them to tenths or hundredths first.

A $\frac{7}{10} =$ ______ $\frac{85}{100} =$ ______ $\frac{1}{5} =$ ______ $\frac{2}{5} =$ ______

B $\frac{2}{10} =$ ______ $\frac{1}{4} =$ ______ $\frac{17}{100} =$ ______ $\frac{9}{100} =$ ______

C $\frac{3}{20} =$ ______ $\frac{48}{50} =$ ______ $\frac{3}{4} =$ ______ $\frac{1}{25} =$ ______

D $\frac{93}{100} =$ ______ $\frac{10}{20} =$ ______ $\frac{12}{25} =$ ______ $\frac{4}{5} =$ ______

E $\frac{33}{50} =$ ______ $\frac{19}{25} =$ ______ $\frac{11}{20} =$ ______ $\frac{21}{25} =$ ______

Name ______________________________ Date ______________

Decimal Review

Write a fraction and a decimal to describe the shaded part of each picture.

A

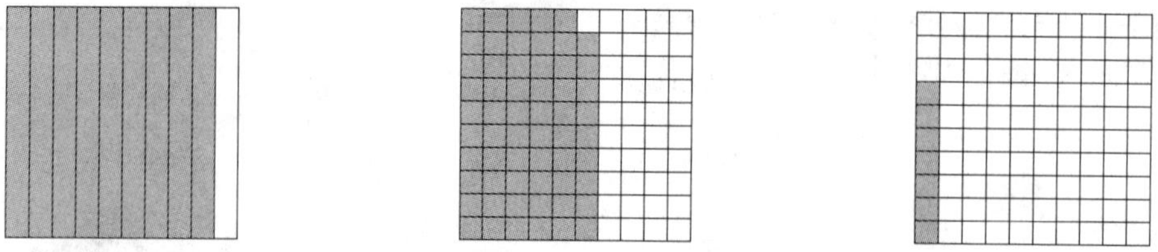

__________ __________ __________

Write the matching fractions.

B 0.7 ________ 0.9 ________ 0.27 ________ 0.03 ________

Write the matching decimals.

C $\frac{6}{10}$ ________ $\frac{25}{100}$ ________ $\frac{6}{100}$ ________ $9\frac{1}{10}$ ________

Write the numbers in order from the least to the greatest.

D

0.68	________
0.8	________
0.86	________
0.6	________

E

0.49	________
0.3	________
0.05	________
0.7	________

F

1.2	________
1.02	________
1.12	________
1.21	________

Round to the nearest whole number.

G 3.5 ________ 5.7 ________ 2.19 ________ 9.58 ________

Round to the nearest tenth.

H 8.76 ________ 9.03 ________ 12.65 ________ 14.29 ________

Name ______________________________ Date ______________

More Decimal Review

Add or subtract.

A	0.8 + 0.9	1.5 + 9.3	3.4 − 0.4	7.6 − 1.8	9.8 + 7.7
B	12.42 − 6.13	9.06 − 8.04	15.61 − 8.33	20.05 − 6.24	13.42 − 9.65
C	7.5 + 8.43	4.7 + 6.52	18.6 − 9.43	24.5 − 12.86	30.7 + 9.83
D	52.62 + 18.9	14.03 + 25.7	26.34 − 16.8	39.06 − 9.4	17.35 + 18.5
E	6.3 2.85 + 4.03	12.9 7.5 + 15.06	16.37 4.28 + 3.9	5.3 7.64 + 8.4	14.87 12.09 + 6.3

Solve.

F Choose an addition problem from above. Write a subtraction problem that has the same answer.

____

____

G Choose a subtraction problem from above. Write an addition problem that has the same answer.

____

____

Answer Key

Identifying Parts of a Whole (page 4)

A. $\frac{1}{2}$, $\frac{1}{2}$
B. $\frac{2}{5}$, $\frac{3}{5}$
C. $\frac{2}{3}$, $\frac{1}{3}$
D. $\frac{3}{8}$, $\frac{5}{8}$
E. $\frac{5}{6}$, $\frac{1}{6}$
F. $\frac{7}{10}$, $\frac{3}{10}$
G. $\frac{5}{8}$, $\frac{3}{8}$
H. $\frac{8}{10}$, $\frac{2}{10}$

Identifying Parts of a Group (page 5)

A. $\frac{1}{3}$
B. $\frac{4}{5}$
C. $\frac{3}{4}$
D. $\frac{5}{9}$
E. $\frac{2}{7}$
F. $\frac{5}{6}$
G. $\frac{3}{8}$
H. $\frac{2}{3}$
I. $\frac{7}{10}$
J. $\frac{4}{5}$, $\frac{1}{5}$

Comparing Fractions—Like Denominators (page 6)

A–D: Bars should be colored accordingly.
A. $\frac{2}{3}$
B. $\frac{3}{4}$
C. $\frac{4}{5}$
D. $\frac{7}{8}$
E. $<$ $>$ $>$ $<$
F. $<$ $<$ $<$ $>$
G. Kate
H. Brett
I. The fraction with the larger numerator is greater.

Comparing Fractions—Like Numerators (page 7)

A. $<$ $<$ $>$ $>$
B. $>$ $>$ $<$ $<$
C. $<$ $<$ $>$ $>$
D. Lee
E. Evan
F. most popular—pepperoni
least popular—ham

Ordering Fractions (page 8)

A. $\frac{1}{5}$, $\frac{2}{5}$, $\frac{3}{5}$, $\frac{4}{5}$
B. $\frac{2}{6}$, $\frac{3}{6}$, $\frac{4}{6}$, $\frac{5}{6}$
C. $\frac{2}{10}$, $\frac{5}{10}$, $\frac{7}{10}$, $\frac{9}{10}$
D. $\frac{1}{8}$, $\frac{3}{8}$, $\frac{5}{8}$, $\frac{6}{8}$
E. $\frac{1}{9}$, $\frac{1}{8}$, $\frac{1}{6}$, $\frac{1}{5}$
F. $\frac{2}{10}$, $\frac{2}{7}$, $\frac{2}{4}$, $\frac{2}{3}$
G. $\frac{5}{12}$, $\frac{5}{11}$, $\frac{5}{8}$, $\frac{5}{7}$
H. $\frac{3}{12}$, $\frac{3}{10}$, $\frac{3}{8}$, $\frac{3}{4}$
I. $\frac{4}{9}$, $\frac{4}{8}$, $\frac{4}{6}$, $\frac{4}{5}$
J. $\frac{6}{12}$, $\frac{6}{10}$, $\frac{6}{8}$, $\frac{6}{7}$

Equivalent Fractions (page 9)

A. $\frac{3}{6}$
B. $\frac{2}{6}$
C. $\frac{2}{8}$
D. $\frac{6}{10}$
E. $\frac{4}{8}$
F. $\frac{4}{6}$
G. $\frac{4}{8}$, $\frac{6}{8}$
H. $\frac{3}{9}$, $\frac{6}{9}$
I. $\frac{4}{4}$, $\frac{6}{6}$

Finding Equivalent Fractions (page 10)

A. $\frac{2}{6}$, $\frac{4}{8}$, $\frac{3}{12}$
B. $\frac{2}{5}$, $\frac{1}{2}$, $\frac{1}{3}$
C. $\frac{5}{10}$, $\frac{6}{12}$, $\frac{4}{12}$
D. $\frac{3}{4}$, $\frac{1}{4}$, $\frac{1}{4}$
E. $\frac{3}{4}$, $\frac{10}{15}$, $\frac{3}{5}$

Simplest Form (page 11)

A. $\frac{2}{4}$ $\frac{1}{2}$
B. $\frac{9}{12}$ $\frac{3}{4}$
C. $\frac{2}{8}$ $\frac{1}{4}$
D. $\frac{5}{10}$ $\frac{1}{2}$
E. $\frac{3}{6}$ $\frac{1}{2}$
F. $\frac{6}{9}$ $\frac{2}{3}$
G. $\frac{1}{3}$, $\frac{1}{3}$, $\frac{4}{5}$, $\frac{1}{4}$
H. $\frac{1}{2}$, $\frac{5}{6}$, $\frac{1}{5}$, $\frac{3}{4}$
I. $\frac{1}{3}$, $\frac{3}{5}$, $\frac{1}{2}$, $\frac{2}{3}$

Fractional Part of a Number (page 12)

Balloons should be circled to form 2 equal groups.

A. 2 balloons colored; 2
4 balloons colored; 4
3 balloons colored; 3

Balloons should be circled to form 3 equal groups.

B. 1 balloon colored; 1
3 balloons colored; 3
4 balloons colored; 4

Balloons should be circled to form 4 equal groups.

C. 2 balloons colored; 2
1 balloon colored; 1
3 balloons colored; 3

Parts of a Number (page 13)

A. 10, 8, 2
B. 6, 6, 8
C. 16, 10, 6
D. 10, 9, 8
E. 14, 15, 12

Improper Fractions and Mixed Numbers (page 14)

A. $\frac{3}{2}$, $1\frac{1}{2}$
B. $\frac{15}{4}$, $3\frac{3}{4}$
C. $\frac{11}{2}$, $5\frac{1}{2}$
D. $\frac{13}{8}$, $1\frac{5}{8}$
E. $\frac{11}{3}$, $3\frac{2}{3}$
F. $\frac{13}{6}$, $2\frac{1}{6}$
G. $\frac{16}{2}$, 8
H. $\frac{12}{4}$, 3
I. $\frac{9}{3}$, 3

Changing Improper Fractions to Mixed Numbers (page 15)

A. $1\frac{1}{2}$, 2, $1\frac{2}{5}$, $2\frac{1}{4}$
B. 3, $5\frac{1}{2}$, $3\frac{3}{4}$, $1\frac{1}{7}$
C. 2, $2\frac{2}{5}$, $1\frac{3}{8}$, 5
D. 8
E. 6

Changing Mixed Numbers to Improper Fractions (page 16)

A. $\frac{3}{2}$, $\frac{7}{4}$, $\frac{13}{8}$, $\frac{4}{3}$
B. $\frac{5}{2}$, $\frac{13}{4}$, $\frac{29}{5}$, $\frac{19}{8}$
C. $\frac{11}{3}$, $\frac{61}{10}$, $\frac{13}{3}$, $\frac{19}{5}$
D. Yes, there are enough pieces for 21 students because 30 pieces were cut.
E. 7

Adding and Subtracting Fractions with Like Denominators (page 17)

A. 3 parts colored, $\frac{3}{4}$; 2 parts colored, $\frac{2}{3}$; 4 parts colored, $\frac{4}{5}$
B. 5 parts colored, $\frac{5}{6}$; 3 parts colored, $\frac{3}{5}$; 5 parts colored, $\frac{5}{8}$
C. 2 parts crossed out, $\frac{2}{5}$; 1 part crossed out, $\frac{1}{3}$; 1 part crossed out, $\frac{2}{4}$ or $\frac{1}{2}$
D. 4 parts crossed out, $\frac{3}{8}$; 4 parts crossed out, $\frac{1}{6}$; 7 parts crossed out, $\frac{2}{10}$ or $\frac{1}{5}$
E. $\frac{5}{7}$, $\frac{2}{9}$, $\frac{7}{10}$, $\frac{2}{6}$ or $\frac{1}{3}$
F. $\frac{7}{9}$, $\frac{1}{10}$, $\frac{6}{7}$, $\frac{9}{10}$

Adding Fractions with Unlike Denominators (page 18)

A. $\frac{5}{6}$
B. $\frac{5}{12}$
C. $\frac{9}{10}$
D. $\frac{7}{8}$
E. $\frac{7}{8}$
F. $\frac{8}{9}$
G. $\frac{7}{10}$

H. $\frac{7}{12}$

I. $\frac{11}{12}$

J. $\frac{15}{16}$

K. $\frac{7}{8}$

L. $\frac{13}{20}$

Subtracting Fractions with Unlike Denominators (page 19)

A. $\frac{5}{9}$

B. $\frac{5}{12}$

C. $\frac{1}{6}$

D. $\frac{3}{20}$

E. $\frac{3}{8}$

F. $\frac{5}{21}$

G. $\frac{1}{12}$

H. $\frac{1}{8}$

I. $\frac{7}{20}$

J. $\frac{1}{9}$

K. $\frac{5}{12}$

L. $\frac{4}{9}$

Adding and Subtracting Fractions (page 20)

A. $\frac{1}{2}$

B. $\frac{4}{5}$

C. $\frac{1}{4}$

D. $\frac{1}{4}$

E. $\frac{5}{6}$

F. $\frac{1}{2}$

G. $\frac{1}{3}$

H. $\frac{1}{6}$

I. $\frac{3}{5}$

J. $\frac{5}{8}$

K. $\frac{1}{3}$

L. $\frac{3}{4}$

M. $\frac{1}{5}$

N. $\frac{1}{2}$

O. 1

P. $\frac{5}{6}$

Adding Mixed Numbers (page 21)

A. $4\frac{3}{5}$, $9\frac{2}{3}$, $5\frac{7}{8}$, 11

B. $6\frac{1}{3}$, $10\frac{4}{7}$, $11\frac{1}{2}$, $12\frac{1}{8}$

C. $14\frac{1}{3}$, $7\frac{2}{3}$, $8\frac{5}{9}$, $16\frac{1}{5}$

Subtracting Mixed Numbers (page 22)

A. $2\frac{4}{9}$, 1, $\frac{1}{3}$, $3\frac{7}{10}$

B. $3\frac{1}{2}$, $\frac{2}{5}$, $1\frac{1}{4}$, $3\frac{1}{12}$

C. $5\frac{2}{5}$, $3\frac{1}{3}$, $4\frac{1}{6}$, $6\frac{1}{2}$

Problem Solving with Fractions (page 23)

A. 5

B. 10

C. 12

D. 5

E. 24

F. 8

G. $\frac{1}{4}$

H. $\frac{1}{6}$

Fraction Review (page 24)

A. $\frac{3}{4}$

B. $\frac{5}{9}$

C. $> <$ $< >$

D. $\frac{3}{6}$, $\frac{2}{8}$, $\frac{2}{6}$, $\frac{4}{8}$

E. $\frac{4}{6}$, $\frac{1}{2}$, $\frac{3}{4}$, $\frac{4}{5}$

F. $\frac{1}{3}$, $\frac{1}{2}$, $\frac{1}{4}$, $\frac{3}{4}$

G. 3

H. 2

I. 2

J. 6

K. $\frac{5}{2}$, $2\frac{1}{2}$

L. $\frac{11}{3}$, $3\frac{2}{3}$

More Fraction Review (page 25)

A. $\frac{1}{2}$, $\frac{2}{3}$, $\frac{2}{3}$

B. $1\frac{1}{3}$, 1, $1\frac{4}{5}$

C. $\frac{1}{2}$, $\frac{3}{4}$, $\frac{5}{8}$

D. $6\frac{4}{5}$, $8\frac{1}{5}$, $7\frac{1}{8}$, 11

E. $\frac{3}{8}$, $\frac{2}{5}$, $\frac{2}{3}$

F. $\frac{1}{12}$, $\frac{1}{8}$, $\frac{3}{4}$

G. $\frac{1}{4}$

H. Suggested answer: Since $\frac{1}{4}$ of 8 pieces is 2, Sarah ate 2 pieces. Kyle ate 4. That left 2 pieces, or $\frac{1}{4}$ of 8.

Identifying Tenths (page 26)

A. $\frac{3}{10}$, 0.3, three tenths

B. $\frac{2}{10}$, 0.2, two tenths

C. $\frac{6}{10}$, 0.6, six tenths

D. $\frac{4}{10}$, 0.4, four tenths

E. $\frac{8}{10}$, 0.8, eight tenths

F. $\frac{5}{10}$, 0.5, five tenths

G. $\frac{9}{10}$, 0.9, nine tenths

Identifying Hundredths (page 27)

A. $\frac{15}{100}$, 0.15
$\frac{72}{100}$, 0.72
$\frac{50}{100}$, 0.50

B. $\frac{7}{100}$, 0.07
$\frac{40}{100}$, 0.40
$\frac{8}{100}$, 0.08

C. 55 squares should be colored

D. 80 squares should be colored

E. 6 squares should be colored

Decimals Greater Than 1 (page 28)

A. $1\frac{3}{10}$, 1.3
$1\frac{7}{10}$, 1.7
$2\frac{9}{10}$, 2.9

B. $1\frac{47}{100}$, 1.47
$2\frac{9}{100}$, 2.09
$3\frac{65}{100}$, 3.65

C. 3.16, 7.5, 9.89, 4.1

Matching Fractions and Decimals (page 29)

A. $\frac{3}{10}$, $\frac{4}{10}$, $\frac{9}{10}$, $\frac{17}{100}$

B. $\frac{1}{100}$, $\frac{83}{100}$, $\frac{3}{100}$, $\frac{1}{10}$

C. 0.2, 0.7, 0.95, 0.61

D. 0.85, 0.02, 0.46, 0.05

E. $7\frac{7}{10}$, $12\frac{3}{100}$, $5\frac{3}{10}$, $11\frac{4}{10}$

F. $8\frac{9}{100}$, $4\frac{59}{100}$, $6\frac{2}{100}$, $15\frac{1}{100}$

G. 3.1, 5.7, 10.35, 4.65

H. 9.09, 14.3, 17.1, 7.01

Comparing Tenths (page 30)

For A–D, pictures should be colored accordingly.

A. 0.7

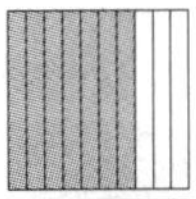

B. 0.9

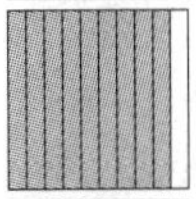

C. 1.6

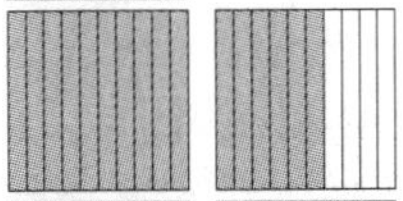

D. 1.9

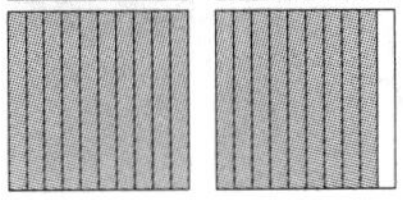

E. $< >$ $> <$

F. $< <$ $> <$

G. Kelly

H. Janet

Ordering Tenths (page 31)

A. 0.2

B. 0.3

C. 0.5

D. 0.8

E. 0.9

F. 1.2

G. 1.3

H. 1.6

I. 0.2, 0.3, 0.7, 0.9

J. 1.3, 1.5, 1.6, 1.8

K. 0.6, 0.9, 1.3, 2.0

L. 0.5, 0.6, 0.7, 0.8

M. 5.5, 5.6, 5.7, 5.8

N. 7.9, 8.0, 8.1, 8.2

O. 4.5, 4.4, 4.3, 4.2

P. 8.2, 8.1, 8.0, 7.9

Q. 12.0, 11.9, 11.8, 11.7

Comparing and Ordering Hundredths (page 32)

A. > 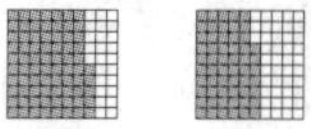

B. < 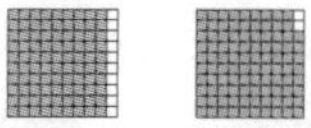

C. = 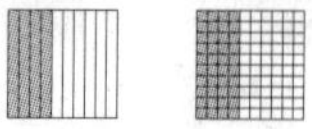

D. < 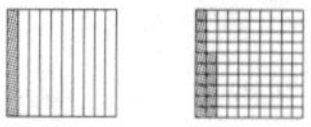

E. = 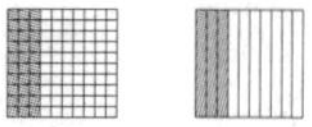

F. > 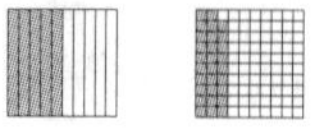

G. < = <>

H. > < <>

I. > < = <

Comparing and Ordering Fractions and Decimals (page 33)

A. greatest—0.32; least—0.23
B. greatest—2 5/10; least—2.1
C. greatest—4.6; least—4 40/100
D. greatest—9.8; least—9 7/10
E. 0.09, 0.7, 0.79, 0.9
F. 1.08, 1.2, 1.28, 1.5
G. 16/100, 0.3, 0.39, 4/10
H. 0.2
I. 0.35
J. 0.75
K. 0.9
L. 1.25
M. 1.4
N. 1.7
O. 1.82

Rounding Decimals (page 34)

A. 8, 4, 9, 8
B. 3, 10, 13, 11
C. 6, 16, 15, 3
D. 1.4, 6.1, 5.8, 9.5
E. 20.9, 15.4, 26.6, 11.1
F. 0.6, 17.0, 40.1, 32.8
G. 5
H. 6.3

Adding Tenths (page 35)

A. 0.6, 1.7, 3.8, 2.9, 15.0
B. 9.2, 29.1, 51.0, 36.8, 39.2
C. 30.4, 33.8, 68.6, 25.4, 40.0
D. 5.3, 0.9, 2.2
E. 1.9, 3.0, 4.7
F. 6.8, 7.9, 10.6

Subtracting Tenths (page 36)

A. 5.1, 3.2, 2.4, 3.7, 16.5
B. 16.7, 9.4, 29.9, 17.2, 49.9
C. 22.3, 14.7, 18.1, 2.8, 34.8
D. 5.7
E. 10.8

Adding and Subtracting Tenths (page 37)

A. 6.8, 7.8, 7.9, 19.0, 35.7
B. 11.0, 14.8, 11.5, 18.9, 19.1
C. 36.8, 42.1, 45.5, 21.6, 31.0
D. 6.3, 45.1, 53.1, 28.1, 42.7
E. 10.3, 24.1, 8.5, 42.0, 22.5
F–G. Problems will vary.

Adding Hundredths (page 38)

A. 8.84, 10.26, 6.17, 4.74, 1.53
B. 12.80, 2.81, 8.24, 9.62, 16.92
C. 38.17, 29.25, 20.58, 20.54, 41.73
D. 27.25
E. 56.70

Subtracting Hundredths (page 39)

A. 4.52, 2.21, 8.36, 17.05, 21.16
B. 26.22, 23.56, 23.12, 8.74, 62.17
C. 14.57, 22.55, 11.63, 35.04, 40.57
D. 5.95
E. 1.15

Problem Solving with Decimals (page 40)

A. 22.89 pounds
B. 73.4 pounds
C. 32.41 pounds
D. Ruff, Scooter
E. Ruff
F. He is less; 10 ½ = 10.5 (It is halfway between 10 and 11.) 10.5 equals 10.50, and that is greater than 10.39.
G. more; 43.1 pounds

Decimals, Fractions, and Money (page 41)

Coin	Number of Cents	Fraction of a Dollar	Decimal Amount
penny	1¢	$\frac{1}{100}$	$0.01
nickel	5¢	$\frac{5}{100}$	$0.05
dime	10¢	$\frac{10}{100}$	$0.10
quarter	25¢	$\frac{25}{100}$	$0.25
half dollar	50¢	$\frac{50}{100}$	$0.50

Coin	Number of Cents	Fraction of a Dollar	Decimal Amount
8 pennies	8¢	$\frac{8}{100}$	$0.08
5 nickels	25¢	$\frac{25}{100}$	$0.25
7 dimes	70¢	$\frac{70}{100}$	$0.70
3 quarters	75¢	$\frac{75}{100}$	$0.75
2 half dollars	100¢	$\frac{100}{100}$	$1.00

A. 83/100
B. 49/100
C. 19/100
D. 37/100
E. 21/100
F. 99/100
G. Julie is right. Seventy pennies equal 70/100 of a dollar, the same as 7/10.

Decimal Patterns (page 42)

A. 1.0, 1.2, 1.4, 1.6 0.2
B. 0.25, 0.30, 0.35, 0.40 0.05
C. 0.9, 0.8, 0.7, 0.6 0.1
D. 0.56, 0.45, 0.34, 0.23 0.11
E. 3.2 miles
F. 10.9 centimeters
G. 2.25 pounds

Fraction and Decimal Equivalents (page 43)

A. 0.7, 0.85, 0.2 or 0.20, 0.4 or 0.40
B. 0.2, 0.25, 0.17, 0.09
C. 0.15, 0.96, 0.75, 0.04
D. 0.93, 0.50, 0.48, 0.8 or 0.80
E. 0.66, 0.76, 0.55, 0.84

Decimal Review (page 44)

A. 8/10, 0.8; 59/100, 0.59; 7/100, 0.07
B. 7/10, 9/10, 27/100, 3/100
C. 0.6, 0.25, 0.06, 9.1
D. 0.6, 0.68, 0.8, 0.86
E. 0.05, 0.3, 0.49, 0.7
F. 1.02, 1.12, 1.2, 1.21
G. 4, 6, 2, 10
H. 8.8, 9.0, 12.7, 14.3

More Decimal Review (page 45)

A. 1.7, 10.8, 3.0, 5.8, 17.5
B. 6.29, 1.02, 7.28, 13.81, 3.77
C. 15.93, 11.22, 9.17, 11.64, 40.53
D. 71.52, 39.73, 9.54, 29.66, 35.85
E. 13.18, 35.46, 24.55, 21.34, 33.26
F–G. Problems will vary.